I0032357

Alexander H. Dana

Ethical and Physiological Inquiries

Chiefly relative to subjects of popular interest

Alexander H. Dana

Ethical and Physiological Inquiries
Chiefly relative to subjects of popular interest

ISBN/EAN: 9783337062965

Printed in Europe, USA, Canada, Australia, Japan

Cover: Foto ©berggeist007 / pixelio.de

More available books at **www.hansebooks.com**

AND

Physiological Inquiries,

CHIEFLY RELATIVE TO

SUBJECTS OF POPULAR INTEREST.

BY

A. H. DANA.

New-York:

CHARLES SCRIBNER, 124 GRAND STREET.

LONDON: SAMPSON LOW, SON & CO.

———

1862.

Entered according to Act of Congress, in the year 1862, by

A . H . D A N A ,

In the Clerk's Office of the District Court of the United States, for the Southern
District of New-York.

Preface.

--- • • • ---

THE following Essays are fragments—not that the thoughts expressed in them are merely of first impression; on the contrary, what I have written has been well considered. But in respect to most of the Essays, they are suggestions rather than conclusions, which it was my intention, when the requisite leisure could be obtained, to elaborate and amplify into treatises more extended in argument and illustration.

Having now little hope of the leisure or health required for carrying out my original plan, I consign to the public these fragmentary thoughts, such as they are, and will only add that they comprise but a small part of the reflections which have been the result of the study and meditation of many years of a life which has not been idly spent.

INDEX.

viii

The following letters, from two of our most eminent scholars and divines, will, I trust, not be deemed an inappropriate introduction to this volume.

In making such use of them, I have in view only the more effectual accomplishment of the object for which these Essays are placed before the public, viz., the furnishing of useful suggestions to those who can be induced to **think.**

But it is necessary **first** to get attention, and these distinguished, but perhaps too partial friends, will **be** listened to with a respect which would not be **conceded to an** author comparatively unknown :

FROM REV. HORATIO POTTER, D. D., BISHOP OF THE DIOCESE OF N. Y.

My Dear Sir—I have to thank you for your kindness in allowing me to look at the manuscript of your "Ethical and Physiological Inquiries," and to express my regret that **very** urgent engagements put it out of **my** power **to peruse and reperuse the whole, as** I could wish **to do.** But I have **seen enough to make me feel desirous to see it in** print. The portions examined—*Influence of Great Men, Hereditary Character, Lawyers, etc.*—interested me much. They furnish, indeed, a kind of reading which always has a charm for me, and I much mistake the taste of a certain class of readers, if these curious observations do not meet with a friendly reception.

It is very delightful to me, my dear **sir, that, after the lapse of more** years **than** I **have courage** to count **up, my old college** friend **should** return to me with pleasant discourse, **which** recalls so vividly the **glowing**, youthful talk of former days. **I see even now, as** if present, the dear **little** room in college, and the two genial students, filling up the intervals **of severer** study with high debate about poetry, and history, and science, **and philosophy,** and loving each other as much as they loved knowledge. Precious, happy days! Blessed beginning of pleasures which have since been flowing through all the life! the pleasure of reading and thinking.

I remain, my dear sir,

Ever faithfully yours,

HORATIO POTTER.

FROM REV. R. S. STORRS, Jr., D. D.

A. H. DANA, Esq.,

My Dear Sir—I have read **many** of your essays, as those on *Identity, Necessity, Races, Authorship, Influence of Great Men, Lawyers, Sensuality, Narcotics*, and have been much interested in and instructed by **them**. They strike me in general (without particular criticism **on** either) as containing the fruits of much research and reflection, and the germs of much **more elaborate** treatises. I should be exceedingly **glad** to have them on **hand in the form of a volume**, and I am sure **that many others would also.**   *   *   *   *   *   *   *

Whether it would be possible, at the present time especially, to reach the popular mind, so as to **secure** for them a remunerative circulation, you can decide better than I. **I am very glad to have seen** and read these writings. They have made me know you better, **and** have given me several most pleasant half hours.

Ever truly yours,

R. S. STORRS, Jr.

I.

RACES OF MEN.

THE whole period in which the Greeks had much distinction, viz., from the Persian invasion to the subjugation of Greece by the Macedonians, when their various popular and aristocratic governments became merged in a military monarchy, was little more than 150 years.*

After that period there were some men of renown in war, but they were not the representatives of the national mind, nor had any more influence for the elevation of their country than those conquerors who have been produced in the *East* in later times, whose power has been that of destruction—an unmixed agency of present evil without productiveness of future good.

The artistic and literary character of the Greeks disappeared after the Macedonian conquest ; and even in war the *rudest tribes* subsequent to that time furnished a more warlike soldiery, and more distinguished military chiefs, than the once renowned states of Athens and Sparta.

---

* The Battle of Marathon was........ ....490 B. C.
The Battle of Chæronea was..............................338 "

It would seem to show that the peculiarities of a race depend more on the circumstances under which national character is developed than on hereditary transmission ; yet the extinction of the genius of the Greeks may be sufficiently accounted for by the crushing despotism of the Macedonian conquerors, and the more annihilating effect of Roman oppression.

The rule of the first conquerors was accompanied by many acts of lawless violence, yet still left some remnant of ancient laws and usages. The Roman conquest subverted all that remained of freedom. The worst vices predominated after that time—the literature of the country became a subtle and unmeaning sophistry, and the nation that once exhibited the greatest development of intellectual power that has been seen, until a recent period of modern history, has ever since displayed the opposite extreme, viz., the predominance of sensual passion, and the propensities of an enslaved mind.

In the Greek character, whatever there was of genius in the arts or literature, belonged almost exclusively to the Athenians. The few of foreign birth that attained distinction were either educated at Athens or under the instruction of teachers sent forth from its schools. Other Greek nationalities had but one thing in common with the Athenians, viz., the power of self-government. The degree of freedom was indeed various, but under every form of government some class of the community was independent. Even Athens was not entirely democratic ; the free citizens were probably not more than a **fifth of** the population. It has been argued that the superiority of the Athenian mind was attributable to the foreign intermixture, but this is certainly a mistake—for, although for the purposes of trade, free access was permitted to foreigners, yet no such privileges were conceded as would have made a *permanent* residence in Athens desirable. *Political franchise* or *citizenship belonged only to those* **whose** *parents were* **both** *citizens.* **The** property of foreigners who died in Athens **was** appropriated to public use.

The principal means of acquiring citizenship was military service. This service was the peculiar privilege of citizens, but emergencies often occurred when there was a demand for a larger force **than** could be supplied **by citizens,** especially **in the naval** service. **In most** cases, foreigners and even **slaves,** might be enrolled, **and for** any valuable achievement, might **be rewarded** by a gift of citizenship. **It is** not **known** with certainty **how the de**scendants of residents who **were** natives, **but not** citizens, were regarded. It may **be inferred, how**ever, from all that we have of historical data, that no one could claim the privileges of a citizen unless both his parents were citizens. This must **have** been an effectual prohibition of marriage **between citizens and foreigners.**

1*

It is most likely that the increase of citizens was chiefly from slaves. Though this might, to a certain extent, be called a foreign element, as the slaves were of alien extraction, it can hardly be deemed an improvement in the national stock. It was better, indeed, that the slaves should be free, but the nations from whom those slaves had been taken were inferior to the Athenians—a large proportion were *Asiatic*—and, besides the habits acquired by slavery, must have been little suited to a preparation for the duties of citizens.

The most remarkable intermixture in European nations was that of the Normans and Anglo-Saxons. The latter were not of a high order in intellectual or moral character ; their distinguishing trait was the love of personal freedom—but the lawlessness produced by the propensity which every man had to do as he pleased, was destructive of the very freedom which was so much prized. The power of the strong prevailed over the weak ; the laws were ineffectual for the protection of the common people ; it had, therefore, become necessary for the lower class to seek the patronage of powerful individuals of the nobility. Even the lesser *Thanes* had to yield a certain fealty to those of higher degree ; a large proportion of the nation was in a state of servitude before the Norman conquest. The Norman tyranny was felt chiefly by the Saxon nobility and large landholders, but the condition of the common

people was improved in two respects ; first, by
the greater authority of the laws, which, though
unequal and in many respects oppressive, were
still enforced for the protection of the common
people ; second, in the religious influence introduced by the Normans, which afforded many aids
to the poorer class.

The *Norman* race was, at the time of the conquest, intellectually superior to the *Saxon*—more
accustomed to the restraint of law, and more habit-
uated to respect for religion. The character of the
two races is strongly contrasted by the systematic
oppression exercised by the one in conformity with
laws, and the tenacity of the other in adhering to
usages which never had authority by legislative
enactment, but which were indelibly impressed upon
the national mind.

It is probable that the middle class, including
those of the lesser Thanes and landholders, who
were not conspicuous enough to become the subjects of Norman rapacity, were the principal assert-
ers of traditional rights. The personal independence of that class was not entirely subverted ;
there were many occasions when they could be useful to their superior lords, and their united power
could not be safely contemned by any Norman baron.

It was this personal independence, this strong
adherence to the Saxon principle of yielding obedience to and claiming rights in accordance with

usages which were supposed to be founded upon voluntary assent, that constituted the characteristic of the Saxon, after the loss of his national liberty. The degree of this tenacity is best proved by the result. The language of the subject race finally obtained the ascendency, as did also that *unwritten law* so long contended for, and which was gradually forced upon the judges by the pertinacity of the common people.

The effect of intermarriage of the two races is of far less account than the counteracting forces which were brought into operation, and which held each other in check. The *slow* progress by which the English character was developed, has contributed, no doubt, to its stability.

The experiment is now being tried in this country, for the first time in the world's history, of the free admission of foreigners to the privileges of citizenship. The result of that experiment cannot now be foreseen with certainty, but upon it depend consequences which cannot be calculated by the most far-seeing sagacity. I am not a *politician*, as that term is commonly understood, viz., as meaning one whose views are limited to the *present* objects sought by either of the great parties of the country.

The speculations which have occupied my mind are rather ethnological than political, and have reference to changes in the condition of this country which will not be fully developed till after the present generation shall have passed away.

No one, perhaps, except those **who are** critically versed in the history of other nations, is aware that there never has been a single precedent of the indiscriminate admission of aliens to the rights of citizenship until it was tried **in this country.** The practice in England was the nearest approach to it ; but the privileges conferred by the English laws consisted only of security of life and liberty, and the right **of** acquiring personal property. A **foreigner** could not become entitled to hold real estate but **by** act of Parliament ; whether the *elective franchise* could **be** acquired **at** all, might admit of **serious** question—certainly nothing **short of an act of** Parliament **would be sufficient.**

In the other states of Europe, previous to **the** first French revolution, although a foreigner **had,** under certain restrictions, protection, as respected personal liberty, and might, to some extent, acquire movable property, yet at **his death, the property** left belonged to the state, **unless he** left heirs **who were** citizens of the **country** where he died.

In France, the *"Loi d'Aubaine"* (forfeiture of the **goods of** an alien) was abolished by the Assembly of 1791, but was restored by the Napoleon code, and remained in force until 1819, when the more enlightened practice of England and the United States was adopted, to the extent of allowing *personal property* of a foreigner to be disposed of by will, **or to** descend to his **heirs,** whether citizens or aliens.

In Athens the *residence* of foreigners was encouraged, but the privileges of a citizen could be obtained only by a decree of **two** successive assemblies **of** the people, which was of rare occurrence, and only in the case of a distinguished person, or **some** one who had rendered great service to the republic. **The** property of foreigners was ordinarily appropriated to public use.

In the time of Demosthenes there were 21,000 free citizens in Athens—10,000 strangers ; the number of slaves is extravagantly stated at 400,000. The enumeration of citizens probably included only males, or those who were entitled to vote, or to sit **in the** Dikasteries. Marriages between foreigners and **citizens** were prohibited ; or, in other words, none were admitted to the privileges of citizens except those whose parents (father and mother) were citizens.

At *Rome* a foreigner could not dispose of his property *by will*—at his death his effects belonged **to** the state, or, under certain circumstances, to his *patron*, that is to say, some Roman citizen whose protection he enjoyed in the city. The right of voting in the assemblies **of** the people belonged only to free citizens of Rome, **and those** to whom by law citizenship had been extended. This was for a long time limited **to** the *Latin States*, who had a common origin with the Romans, and were, in language **and** customs, essentially the same people.

Marriages were virtually prohibited **at** Rome between patricians and plebeians, until the fourth century of the nation, when the Canuleian law was passed. Previous to that time the children of a patrician father and plebeian mother were heirs only of the mother.

The employment of a large number of seamen in the Athenian war made a change in the citizenship at *Syracuse*, by which the popular party acquired strength. Hermocrates and his aristocratic friends were banished—Dionysius, not distinguished by birth or fortune, who belonged to the party of Hermocrates, obtained control of the populace, succeeded in getting the exiles recalled, and then, in appropriating to himself supreme power, which he retained until his death, a period of forty years. He admitted to citizenship a large number of enfranchised slaves, and employed mercenary soldiers in place of the citizens of Syracuse of better class.

At *Carthage*, there was an aristocracy of wealth rather than nobility by birth. The character of the people was commercial, and the soldiers employed in war were chiefly foreign. Little is known of the political rights of the citizens of Carthage, or what constituted citizenship. The government was administered by a Senate, which was a large body of men (how chosen does not appear), by whom an executive, consisting of one hundred commissioners, called the Council of the Elders, was appointed.

II

COMPENSATIONS OF LIFE; OR, EQUIVALENTS IN THE CONDITIONS OF MEN.

It would seem, from the uniformity with which men pursue after certain things, as wealth or power, that these were the greatest sources of happiness, or were in themselves most to be desired of all the objects of human pursuit. The testimony of those who have attained them is wholly opposed to this hypothesis, but the world is still no wiser. How is this to be accounted for ? Probably it is because men will seek the gratification of their ruling passions, whatever may be the result, and even with the strongest reason to expect that it will end in disappointment.

There is no passion so strong as the desire of *power*. Wealth is the means of power, and this is a principal reason why it is sought ; for although men who have strong sensual propensities are usually unscrupulous as to the means of indulgence, yet these are ordinarily satisfied if they have enough for the present object, and exhibit little care and assiduity in laying up a fortune for future use.

It is a common inquiry, what is the specific influence of ambition in the world ? is there a balance of evil or of good in its results ? Enterprise in **business**, indeed energy in all the pursuits of life, **seem** to depend upon it, yet it has been the source of the chief disorders in human society.

Our present inquiry, however, is what it contributes to the happiness of those who are swayed by this passion, assuming **that** they are successful in obtaining their end, so far as success may be predicated of any objects of pursuit in this world. Do they obtain that which, according to popular opinion, is an incident, viz., a condition **of** enjoyment superior to that of other **men ?**

The answer is obvious, but lest it should be taken as a mere speculative theory, let some of the counterbalances be considered.

Who ever attained success that has not **been, to** a greater or less extent, unscrupulous, and by consequence become subject to the rebuke of his own conscience ? Who **has** not, in the pursuit of his object, done things to be repented of, sacrificed the claims of private friendship, and violated laws essential to the well being of society ? Again, success is usually the result of assiduous, untiring pursuit, and this, of itself, involves an exclusion of regard for others—it is, in fact, wholly **selfish.** Where are the friends of such a person ? Dependents **or** hangers-on they may have, but the conditions of friendship are wholly wanting. Lastly, no one has attained such a result but through many trials and disappointments ; evil passions have been aroused which cannot be fully gratified. Even the usurper of power, who has become despotic, has not the power to annihilate public opinion ; he knows that he is hated, and that those who have fallen are more cherished in the remembrance of men. On a lesser scale a man who has acquired high station by persevering effort, has to look back upon a course of life devoted to selfish ends, and which will be unhonored when he comes to be judged by the world ; then there is the uneasiness as to retaining possession of what he has acquired—the jealousy of rivals—the hatred of those who have sought unsuccessfully the same honors.

One illustration so striking as this goes far to prove that all conditions are subject to counterbalances, for if that which, by common consent of mankind, is *most desirable*, is in reality so little to be desired in itself, what shall be thought of inferior objects of pursuit ? All men seem to be discontented with what they have, or to suppose that some other condition would be better. Often this may be a vague, unsettled discontent with what they have, rather than a decided preference for another lot, so that if it should be said to them that their wish should be granted, or in the language of Horace—

"Hinc vos,
Vos hinc mutatis discedite partibus,"

they would be found not prepared to take advantage of it.

To the *invalid* health is the object of supreme desire, and is esteemed the greatest of human blessings, yet could we know intimately the history of the man who has had no experience of bodily ailments, he would probably hesitate to make an exchange of conditions. He might find unchecked sensuality a pursuit of self gratification, with no kindly sympathy for any human suffering—no generous regard for the welfare of others—no sense of responsibility to a Supreme and Holy Being ; or if, by education and happy association, there should be some sense of moral rectitude and religious sanction, it may, perhaps, be but a mere theory, assented to as a thing of course, but having no vital influence on his thoughts and feelings. Or, lastly, suppose the best phase of character, and that there is even moral and religious sense, so far as it may subsist without the chastening effect of bodily ailments, there may be other trials—indeed, every man has, sooner or later in life, trials of some sort, and in general it may be said that, in the full possession of physical vigor, there is a lack of the reflective habit, of the calm, considerate, conscientious tone of mind which is essential to a sound character, or even to the proper enjoyment of life.

To one struggling under the discomforts of *poverty*, no state seems comparable to that of affluence ; yet let the poor man remember what the apostle says : " Godliness with contentment is great gain."*
Utter destitution is, indeed, a condition of misery ; but to be obscure, to have enough only for the ordinary wants of our nature, and therefore to be without the consequence which riches confer, is not only not inconsistent with enjoyment, but if religious feeling and a contented mind be superadded, is perhaps highest in the scale of happiness.

There is a wrong judgment prevailing in the world on this subject. Religion does not half fulfil its office while it leaves so great an error unrebuked as the common idea that affluence is the chief element of a happy life.

Again : beauty of person is a natural endowment, which is much prized, and certainly a man is commended to the favor of all persons by fine proportions and a pleasing countenance. If we should suppose that to this be added a competence of fortune and high position by family connection, may it not be supposed that to such a one life has a charm that is denied to others ? The houses of opulent and fashionable families are open to him— he is everywhere greeted with a smile of welcome—

---

* I think this precept is to be understood, that a bare sufficiency for our physical comfort is all that is to be desired ; less than this is a state of actual suffering ; but most of the evils which men suppose to be incident to the want of riches, consist in he privation of means for the gratification of selfish propensities.

even the scenes of nature should wear a lovelier aspect when all the conditions of his life conduce so much to self-complacency.

But let us look a little farther. A vain and frivolous conceit may be the result of flattery, or there may be an utter abandonment to sensual pleasure, which is presented to him under circumstances of great enticement and facility of gratification—a vitiated constitution, ailments never to be recovered from, premature prostration of mind and body, perhaps an early death—these are some of the incidents of unchecked license in the course of pleasure. Or, suppose that by fortunate association, or a greater degree of worldly wisdom than falls to the lot of most men, his enjoyment of pleasure shall have been so moderated that health is not impaired, yet it may be that a selfish, unsympathetic temper may have been developed ; it need not be the malignant passion of a *Commodus*, but the effeminate, self-conceited, ignoble propensities of an *Elagabalus*. One thing is certain, that the tendency of such a condition, enviable as it may be thought to be, is to oppress and prevent the development of those qualities upon which, after all, true happiness depends. An artificial existence takes the place of the natural. There may be social feeling, but it is a heartless conviviality—the mere buoyancy of a physical elasticity. No kindly sympathy was ever yet generated without some discipline of the soul by trials, or by the influence of religion on the heart; but this last is uncongenial with the pampered child of fortune, and almost by necessity disjoined by an immeasurable interval.

It deserves to be considered, also, that when trials do come—and no man is wholly exempt—it will be the more difficult to bear them, for the very reason that there has never been humiliation of spirit, or the repose of a mind accustomed to fall back upon its own resources. Frantic agitation—a breaking of the heart, unalleviated by the softening influence of well-tried friendship—is not this often the experience of those whom the world envies and adulates?

If, however, it could be understood by all what there is of comparative comfort in their own particular condition, and what is the *real* extent of enjoyment in the condition of those seemingly more favored, there would be a great advance in true knowledge. We are strangely imposed upon by mere outward show. The inmate of a luxurious carriage, rolling along the street, an object of envy to her pedestrian sisterhood, perhaps derives from this invidious homage her chief satisfaction; and if, on the other hand, she were assured that all the lookers-on were contented to be without the superfluity which constitutes her distinction, and that they were satisfied with the healthful glow imparted by the natural use of their own limbs, her expensive luxury would lose its charm.

Burns has described the state of the poverty-stricken laborer as having some equivalent :

> " Ye maist wad think a wee touch langer
> An' they maun starve with cauld and hunger,
> But how it comes I never kenned yet,
> They're maistly wonderfu' contented ;
> An' buirdly chiels and clever hizzies
> Are bred in sic a way as this is."

Dr. Kane relates that he was much impressed by the glee and merry laughter of a party of Esquimaux children, who had taken advantage of a moderation of temperature to sport in the open air, though the cold was still such as in our latitude would have been deemed intolerable. "Strange," was his reflection, " that these famine-pinched wanderers of the ice should rejoice in sports and playthings, like the children of our own smiling sky. How strange is this joyous merriment under the desolate shadow of these jagged ice-cliffs."

III.

IDENTITY.

A TREE is the same when it is young and old, and even when it has lost its branches and is but a decaying trunk. Its locality or fixedness in a particular place forms a part of our idea ; observation of it in its various changes is also of some effect in keeping the identity in our minds, but so long as the *outward form* remains substantially the same we recognize the tree in all its variations of growth and decay.

A living organized body constitutes the identity of an animal. The idea of a man (says Locke[*]) is of an animal of such a certain form that we should have that idea of *any* creature having *our* shape, although destitute of reason, and that we could not have the same idea of an animal of any other shape, though it should have reason and speech.

The outward appearance, it is true, is *the man* that we have the idea of in thinking of a man, but there is more than *shape*—there is the expression of character in the face and bodily motion. If this expression should be *totally* changed it would no

---

[*] " Locke on the Understanding," b. 2, c. 27.

longer be the same **person,** although **the** *figure* should remain the same.

The indications of character constitute essentially **the individual,** and if these could be transferred **to** another body, *and this be known by us,* we **should** recognize the identity. The recollection of a man's character, as displayed **in his conversation** and **acts,** makes **the** *Homoousion,* **as the** reminiscence of our thoughts makes the identity **of** our **own** being.

The case **supposed by Locke, that if we** were sure that the soul of ***Heliogabalus*** was **in a hog, we** should not, **therefore, suppose the hog to be Helio-**gabalus, does **not at all decide the question.**

In the first **place, we** cannot **conceive of** character except **by** some *external* expression, and hence we could have no idea of a human soul in a brute animal, which has no expression, nor could we identify the soul even **in** another human body which should be without features **or** other physiognomic instrumentality **for expressing the qualities of** the **soul.**

But, second, the character **of every man is** so intimately **connected with his** corporeal structure, so much **derived from and** dependent on *its* peculiarities, that **the** *whole* **character could not** exist **in any** other structure, and could not, **of** course, be **recognized by us as** entirely identical.

This limitation, it will be understood, is applicable **to** the hypothetical case which I have assumed in a former remark, as to the probable identity of an individual, even if a change of body was possible.

In a *future* state of existence it seems to me that the *same **body*** is necessary for perfect identification, but **we may** suppose a prescience *then* so clear that **the soul may be** discerned by its intrinsic constituents **without the** necessity of any outward indications.

Personal identity (that is, **a man's** identity in his own consciousness) has been **made by some** writers to depend on *memory,* and a doubt **has been** raised whether a man (*i. e.* the living being) **is at** all times the same.* It is said, on the one hand, **not to be** the same, because the consciousness of his **existence at any two moments is not the** same individual action—*i. e.,* **not the same** consciousness, but different **successive consciousnesses. The fallacy of** this **has been exposed** by Butler. **Consciousness** (he says) *presupposes* and does not *constitute* personal identity. The idea of **this identity** arises upon **comparing our** existence at different **periods of** time, **as we have the idea of** similitude **by** comparing triangles.

The consciousness here referred to **is** a perception that the thinking, **sentient** being, **is the same now**

* Stated by Locke thus : " Whether the same person be the same identical substance ;" and he defines *"person"* to be " a thinking, intelligent being." Which proposition (says Butler) is asking whether the same rational being is the same substance—which needs no answer, because *being* and *substance* stand here for the ame idea. See " Butler's Diss. on **Pers.** Ident."

as formerly—which is equivalent to the knowledge we have from observation that any other object is the same now as when contemplated before.

It appears to me that our identity consists in the constitution of our being, by which we are made subject to certain impulses or motives of action, and peculiar modifications of thought and feeling, which together make the individual character ; and this is unchanging through life, though it may be more or less developed in degree—and, so long as the actuating principle exists and is perceived, so long it is the same individual mind or being, both in our own consciousness and as observed by others.

NECESSITY.

As applied to human action, *necessity* must mean the control of the *will*, for there never **was** any question that our actions were *immediately* directed by our volition. Locke explains it with considerable verbal nicety, after the manner of the old scholastic reasoning, that the will is moved by a present *uneasiness*, which, whether it be a sensual want, as **hunger, or** any **other craving,** has a greater determining power than any **good of more** intrinsic value but more **remote.**

This is in substance a determination **of** the **will** by *motives ;* but the question still remains, by what law or agency do motives operate—not as a matter **of** fact what effect a motive actually has—but in **what manner a motive** determines to a particular **action.**

As **a rule of** general application, we would not say that the *same motive* would invariably determine men **to the** same actions ; **but,** speaking of an *individual*, it may be said that **a motive** does act with *certainty*—so that if we **know** *perfectly* the constitution of a man's mind, we should be able to calculate what his **action would be under any given** circumstances. And such is **the** homogeneity **of** different minds, that a proximate degree of **uniformity** is seen in the actions of men, when acted upon **by the** same motives.

This is, in fact, the foundation of **our** reasoning **in** respect to human actions—and from the conformity of motives and actions, *Hume* asserts, that necessity is as fully made out as it is in the material world. Our idea of necessity, in the operations of nature (he says), is an inference from the constant conjunction of certain things ; and this conjunction equally exists in human conduct, **and an** inference may be made with the **like** certainty from *motives* **as to** the voluntary action that will ensue—from character as to the conduct that will be pursued.

Moral influences or motives have not equal certainty in their effect with physical **causes,** so far as appears to our observation, but this is perhaps only the imperfection of our perception. Long experience, and habit of observation, will give a foresight of the conduct of men, that seems to less disciplined **minds** an extraordinary gift, but which in reality is nothing more than superior knowledge of the disposition or natural tendency of the human mind. Knowing this, calculation is very certain as to conduct, and I think it may be laid down as a general proposition that the actions of men are determined with certainty by motives—that is to say, that it is

not by *chance* which way the mind shall decide, **but
that it is absolutely and** invariably determined **to a**
certain action by a particular motive.

Now how much **soever** a motive, as presented **to
different minds, may** vary in its determining **power,
yet to *every mind* it has a certain** adaptedness, pro-
ducing its specific effect, just **as by the** laws of nature one element will, **in** any number of experiments,
unite with **other** elements **in a** uniform **manner,**
well known **beforehand to** the *chemist*.

Edwards explains it, **that the** will **is determined**
by " that motive which, as it stands in the view **of**
the mind, is the strongest ;" that the strength of a
motive is its *tendency* to move the *will*, and that
this tendency is according **to** the apparent good, so
that **" the will always is** as the greatest apparent
good is."

Moral inability he defines **to be the want of inclination, or** the strength **of a contrary** inclination,
which is the **same** thing as **the want of a** sufficient
motive, or the strength of an **opposing motive.** The
will cannot **act contrary to the** strongest motive, for
that would be to suppose the will other **than** it is ;
but there may be **opposition to, or** an endeavor
against **future** inclinations, **or which is the same**
thing, against the **power of** *future motives.* His
argument as **to the** consistency of **moral inability**
with legal requirements, **is** that wickedness would
otherwise **be always excused, as** the greater

the wickedness the greater the inability (that is, want of inclination). But, according to the common sense of mankind, men are held deserving of praise or blame according to their actions, without any abstract question as to the want of inclination or motive.

The opposite theory, viz., what is called the self-determining power of the will, is this—that the soul has a sort of prerogative by which it can act, independent of motives—or, in other words, contrary to the strongest motive. But there is an evident confusion of terms in this proposition. It is apparent that *motive* is used as synonymous with *reason*, and with that explanation it is undeniable that men do not always act in accordance with the strongest motive (*i. e.* reason). This is, indeed, nothing else than to say that the actions of men are often at variance with what they know to be right. Nay, it may be admitted that our nature is so perverse that we sometimes do what we know, or by reflection might know, is not for our interest—assuming, what I am, however, not willing to concede, that interest and right can ever be separated.

But, in such cases, the ruling motive is a perverted desire, which cannot be restrained by a sense of what is just and proper. This seems to me to be the extent of the doctrine of the self-determination of the will, when reduced to exact phraseology.

Bushnell,[*] who insists strongly upon the independence of the will, yet virtually concedes all that in fact legitimately results from dependence upon motives, while he seems to be singularly unconscious of the concession. " If it be true," he says, " that the wrong-doer chooses what to him is the strongest motive, it by no means follows that he acts in the way of a scalebeam, swayed by the heaviest weight, *for the strength of the motive may consciously be derived, in great part, from what his own perversity puts into it."*

The perversity here spoken of, or, as it may be called, wrong-mindedness, is an adaptation of the mind to be acted upon by wrong motives ; nevertheless, the determination of such a mind is as positive a result of motive, wrong though it be, as the wisest decision of a strictly conscientious man is the result of a better motive.

---

[*] " Nature and the Supernatural," c. 2.

V

AUTHORSHIP.

"NEVER write a book," said Talleyrand ; "if you do we shall know all your brains are worth, for as many francs as your book will cost. No man of sense writes books ; the emperor writes no books (this was said before Bonaparte was sent to St. Helena), Socrates never wrote a book."

While a man is living he will perhaps have more repute for wisdom if he be considered *able* to write but does not, in the same way that, as Bacon says (speaking of conversation), "If you dissemble sometimes that you are thought to know, you shall another time be thought to know that you do not." But traditional wisdom is fugitive, and has no lasting influence, except when a chronicler is found to record the original utterances. *Socrates* was immortalized by his disciples, especially by Plato, who reproduced the thoughts expressed by his master in public discourses, or is supposed to have done so.

Dr. Johnson is as famous now for his conversation as he was with his cotemporaries, which we owe to his biographer, Boswell. But in general, what do

2*

we know of men renowned in their own times, but who had no leisure for authorship ?   Some of their sayings may be recorded, but they are for the most part mere historical figures, known by their actions and not by their thoughts.

Want of time may interfere, in many cases, with literary employment, but no one is indifferent to fame as an author.   Literary vanity is the most inordinate of human passions.   Why is it that emulation is so great where the chance is so slight of any lasting distinction, at least as respects the larger number of authors.   The *pleasure* of composition is suggested as the chief incitement to literary labor.   " These," says Buffon, " are the most luxurious and delightful moments of life, which have often enticed me to pass fourteen hours a day at my desk, in a state of transport ; this gratification, more than glory, is my reward."

But this gratification would not exist but for this glory that is in the mind's vision.   The author's pleasure in the happy expression of a thought, or illustration of a truth, is mainly founded upon the effect which it will have upon other minds.   He may be willing to wait patiently for years, till his labors shall be completed, and he shall come forth by publication into association with the world of thought, and be known familiarly, not as the private individual, but by the title of his book.   Yet, while thus toiling, he is continually cheered by the ovation which lies before him in prospect, when men shall come to know the productions upon which so many solitary hours have been employed.

"Solitude would not be endured," as has been said by some one, I do not recollect whom, "if we did not cherish the hope of a social circle in the future, or the imagination of an invisible one in the present."

The author has, in fact, an imaginary audience continually in his view, and it is the supposed sympathy of their minds which creates his chiefest pleasure.

A second source of gratification to an author is more real, viz., the applause of the friends immediately about him; and perhaps this has, after all, a more direct power than the vague expectation of renown among those whom he will never see.

One case there is in which the expression of our thoughts may be in itself an end, and not the preliminary to something beyond. It is when a man has, by severe trials, been forced to seek resources in himself, when he has been oppressed by affliction, and the sympathy of friends has failed, or where the trial has been so overwhelming that human sympathy is valueless—to such a one self-communion is a solace and a source of strength; but especially is it so when the soul has, from the gloom of its own solitariness, found access to God, and the light of the spiritual world has broken in upon it. Thoughts and emotions then crowd upon the mind which it is unwilling to lose its hold of ; it seeks to secure the memory of them by a permanent record ; in time it may become a habit to transcribe all its impressions ; yet who can say that even in such a case there is not a melancholy forethought that this record shall be seen by other eyes, and that sympathy shall be awakened when it shall no longer be of any avail.

To write for the *entertainment* of those who shall read or hear, is very different from writing merely to *convince*. It is one thing to *think correctly* and another to express what we think agreeably. It is a prevalent tendency to seek popular applause rather than practical instruction—to please a critical taste rather than to instruct the understanding. It is true that we are instructed not solely by logical argument ; we are actuated more by impulse than by reason ; we need sound views, presented in a clear and impressive manner, without exaggeration, but most writers and speakers aim at *eclat.* They would prefer to be admired for genius rather than for sound wisdom. So, on the other hand, it is natural for men to be more pleased with the mere conceptions of imagination than with profound thought ; the former task the mind less, and do not interfere so much as the other with the self-complacency of the reader.

The great multiplication of books is sometimes

spoken of **as an** evil, but this is true **only in one**
sense, viz., so far as they are superficial or trashy.
When there is a prevalent propensity to authorship
it is a natural incident that there will **be** a great
deal of frivolous writing, or that the same ideas will
be often reproduced. The effect of the periodical
literature of the present day on the public mind, is,
in my judgment, not favorable to a muscular tone.
Newspapers, literary magazines, and the like, **must,**
in order **to please** popular taste, **consist of light but**
various material. They are suggestive of **many in-**
teresting inquiries, but this is of little **value to any**
but systematic **thinkers, and** generally the effect is
only to distract the mind **and** impair a habit of
consecutive thought. It is true that those who
think much are also great readers, even of this fugitive sort of productions. This is partly for relaxation, but also for the new ideas which may be de-
rived from a heterogeneous source, when there **is**
strong power **of** assimilation. Such reading **may**
be compared to conversation of the gossiping kind,
which may be supposed to be of no great advantage,
yet Sir Walter Scott said he never met with any
man from whom he could **not** learn something in
conversation.* So of books ; it has become almost
a proverbial saying that there is none that does not
contain something that is valuable.† Macaulay's

---

*Fortunes of Nigel, C. 27
† Pliny **made** the remark, **and it has often been repeated** since.

History of **England indicates** an omniverous **habit** of mind ; materials are turned to valuable account which we should **hardly** have supposed would **have** attracted **the attention of** any one but a frivolous **antiquarian—old** songs, obsolete plays, pamphlets, **newspapers,** traditional proverbs—and these not **hunted** up merely for **the occasion, for that** would have been impossible, but **constituting a** familiar lore. *

The chief **value of** Plutarch's **Lives I** consider **to** be the affluence of anecdotes, apophthegms, **and** slight incidents, which were contemned by stately historians, and which **he** was compelled to obtain **by** a process of **filtration,** involving research into a **vast amount** of gossiping productions, **or** what were so esteemed, and **had not sufficient merit to** reach us, except **in the excerpts penned by this most cru**-dite of ancient **writers.**

---

* **Hume,** as an historian, was defective in **this respect. A paragraph from the** *Edinburgh Review* is worth quotation. "His acquaintance with English litera-**ture** was imperfect, in a degree that in our days must be altogether incredible. In **his** day nothing seems to have been called literature except the showy publications that were addressed to the idle and disengaged portion of the public, rather than to the business mind of England. * * In the Parliamentary history, in the state trials, in the law reports, in the pamphlets of the day, at almost all periods of our history of which we have any valuable records, are found masses of thought **to** which, in their **real interest** and importance, and often even in reference to the artistic skill with which arguments of great power are elaborated, the works of our later literature bear no comparison whatever ; and of all these Hume was, except when by bare accident he looked farther than the popular works by which he was **directed to** his authorities, profoundly ignorant."

VI

INFLUENCE OF GREAT MEN.

CELEBRITY is no certain proof of greatness of mind. Several particulars deserve to be noted in determining what constitutes it.

1. Although **energy** of character is derived **from** strength of **passions, yet the accomplishment of** anything **worthy of admiration depends** upon the **control of those passions. In a** man **of** known **capacity,** moderation is always evidence **of** greatness ; **it** shows that he has power over himself, and, therefore, that he is able to observe and take advantage of the indiscretions of other men—to **estimate** deliberately **the present, as compared with the** past or future, without **being imposed upon by** the exaggerated **views** which deceive most **of** mankind. *Talleyrand* was described **in his** early life, by an acute observer, **as** destined to distinction, by reason of the possession of these qualities.* Of *Hampden*

---

* " In his judgment of men he has that indulgence, and in his estimate of events that *sang froid*, and in **all cases** that moderation, which are the genuine marks of **wisdom.** * * He does not **imagine that a** great reputation is to be raised **in a day** ; but he will assuredly accomplish **that** object, for he will never fail to **seize** those opportunities which Fortune so frequently offers to those who do not **violently assail her."** I am indebted for this quotation to the *Dublin University Magazine,* **which does not,** however, give the name of the author.

it is related that he kept himself in reserve in the early part of the contest between the Parliament and king, preparing himself for the crisis which he foresaw was impending. " His carriage," says Clarendon, " throughout this agitation, was with that rare temper and modesty that they who watched him most narrowly, to find some advantage against his person, to make him less resolute in his cause, were compelled to give him a just testimony." He appears to have left to other leaders of the opposition the credit and responsibility, while he calmly waited for the time which should call for his superior ability.

It is an indication of a great mind when a man, conscious of natural power, is patient of the delay which may be interposed to his advancement. A premature attempt might place him in a false position ; at all events, he will not be likely to have the influence which he may expect if he wait his time. It may, indeed, happen that the opportunity may not be offered for the full display of all his capacity. It is an incident of human life that good fortune does not happen to all, yet in the ordinary course every man will sooner or later find the position to which his merit entitles him. Or if, perchance, it should be his lot to go down to death " without his fame," a truly great man will still retain satisfaction in the consciousness of his own worth, for there is more dignity in making small

account of the applause of the world, than in the
possession of the highest popular honors.

2. As a general rule, greatness of mind is gradual
in development—

> "Crescit occulto velut arbor aevo
> Fama Marcelli,"

is a felicitous description of a reputation which has
grown up without being observed—or, rather, of the
development of capacity in unostentatious progres-
siveness, till on some favorable occasion it is exhibited to an admiring world, and then with the
greater *eclat*, in proportion as it had been before
unnoticed. It is, upon the whole, better that a
man should not obtain distinction early, for when
he has once come into the strife of life, "medium
in agmen, in pulverem, in castra," there is no longer opportunity for domestic exercise, and for the
gathering of resources. Thus, those who have by
accidental advantages attracted the notice of the
world early in life, though they may by experience
acquire great readiness and tact, will lack profound
knowledge and comprehensive views. Long preparation is in general essential to the gaining of a
lasting reputation. The maturity of mind, and extended observation belonging to those who have
risen late in life, will insure discretion and practical wisdom, which are important elements of true
greatness, and of permanent influence in the world.

The younger *Pitt* had extraordinary natural gifts,

but in consequence of the *eclat* of his first efforts in
Parliament, having been suddenly called into minis-
terial office, which he continued to hold the rest of
his life, there was observable a deficiency of the resources which a longer pursuit of general literature
would have bestowed.

In contrast with the brilliant early life of Mr.
Pitt was the slow advancement of *Sir Robert Peel.*
At first but little noticed, indeed for a considerable
time hardly thought of as likely to be one of the
great men of his time, his powers expanded in the
faithful discharge of practical duties, till at an advanced period he was acknowledged as entitled to a
place among the ablest English statesmen.*

Marshal Suchet was remarkable for a growth of
capacity in his later years, which, as Napoleon, who
was a great observer of men, said of him, was truly
astonishing.

Bacon has noticed that " some have an over-early
ripeness in their years, which fadeth betimes."
Under which head he mentions Hermogenes, the
rhetorician, who was exceeding subtle, but afterwards became stupid ; the orator Hortensius, of
whom it was said by his rival, " idem manebat nec
idem decebat ;" and Scipio Africanus, whose early
achievements far surpassed all that he afterwards

---

* In one of the sketches that appeared at the time of his death, it was remarked
that his outward appearance corresponded in outward development to his mind.
The sagacious but commonplace countenance of his earlier manhood was marked,
as he advanced in years, by a peculiar expression of refined and playful acuteness.

performed, as was pithily expressed by **Livy**, *"ultima primis cedebant."* **But** this last instance **rather** illustrates that **great genius** is called **into** action by great occasions. Had **another** adversary like Hannibal appeared, it might have been found **that Scipio's** great qualities **were** not impaired. **It was a** misapplication by Bacon, **when** he speaks of Scipio as one of those who take too **high** a strain at first, which they cannot afterwards uphold. **Belisarius might equally have** been supposed **to have** degenerated **from** himself, **if his old** age had **not** been made illustrious **by his** recall from **retirement**, to sustain **the** sinking **fortunes of his** country against the ferocious Bulgarians.

3. *Directness of thought and action* are characteristics of great men. The last has been most observed, especially **in** military men, but the first **is** equally **an** essential **trait,** indeed both are **connected, for as** a man's **thought** is, **so is** his action.

Some eminent men have, **however, lacked fluency or** perspicuity of expression, which has **given** rise to **the idea that** they were of hesitating mind, although decided **in action,** as in the case of Cromwell ; but this may **have** been **mere** want of facility in the use of language, **but** generally we **find** energy **of language** where there **is** energy **of action. The** old veteran, *Suwaroff*, **could say very well** what he meant, and was understood, **as when he wrote to the** Austrian general, Melas, who **was not pleased**

with marching in wet weather : " Fine days were
made," said he, " for women, and fine gentlemen,
and lazy persons."

So Blucher, in his memorable order to the Prussians, announcing that he would lead them against
the French, on the day of the great battle of Waterloo, " We shall beat them, for we must."

In respect to Cromwell, it has recently been
proved, especially by the publication of his letters,
that he was able to express what he meant in very
clear and forcible terms.

4. It might be expected that an incident of greatness of mind would be the power of calling forth
latent ability in other men ; yet this is by no means
uniform, and if it were made a test, would reverse
or materially modify the popular estimation of many
celebrated men.

The highest order of genius is undoubtedly that
which multiplies its own force by the development
of capacity in other men. This requires, in the first
place, profound observation of the character and
natural power of men ; but, in addition to this, a
discriminating judgment as to the manner of eliciting this power. Many men are unconscious themselves of what they are fitted for, and accidental
position sometimes seems to create talents, though
in fact it only calls them into exercise.

In order fully to show what a man is capable of,
he must at some time be put upon his own discre-

tion, and not always be subject to precise and positive instructions, **for** this **is** to keep the mind **in** vassalage to the principal in **his** absence, as **well as** when he is present.

Napoleon had a singular power of making great men, **or,** in other **words,** discovering and eliciting **the** capacity with which they were endowed by nature. His generals and ministers of state had a greatness of their own, which was indeed called forth and disciplined by his peculiar energy, yet was **not a mere** reflection of the ability of their chief ; **hence, in all departments, civil and** military, he was served with a marvellous efficiency, seeming **to** be **like an** infinite multiplication of himself, which astonished the world, and was in fact a harmony of action, and a uniformity of success, that might well seem to result from the forecast of a single mind rather than the divided agency of many.

Frederic the **Great, on the** other hand, **had** no creative power of **this** kind. He allowed little discretion to others, and hence **gave little** opportunity for the development **of** their natural ability. His government was unique and simple, but lacked the splendor as well **as** the intrinsic strength which is derived from the combined action **of many gifted** minds, brought **into** vigorous exercise under the general direction of **a** master-spirit.

Alexander the Great, though he died in early life, **left** many great men, who had been trained **under** him, and whose achievements, after his death, reflected the greatness of their master, and increased the lustre of his fame.

Julius Cæsar, though in executive ability never surpassed, had no faculty of creating, or calling it out in others, except in strict subordination to himself. With the exception of Marc Antony, who, however, derived but little from Cæsar, there was not one distinguished man among his subordinates, either civil or military.

I have considered what constitutes greatness. ˙ I propose now to estimate the influence of great men upon the people to whom they belong.

Military genius may do much, for in war the common soldiers are acted upon by impulse, and the leader may inspire those under him with something of his own resolution. Again, masses of men, when brought together, are wielded by a kind of mechanical agency—discipline gives confidence, each man relies on his associates, and the known ability of the general gives assurance to the whole army.

So it has been often seen that a great general may make a superior soldiery out of indifferent materials. The Thebans became, under Epaminondas, the first military power of Greece, though they had never before been distinguished for prowess. *Hannibal*, out of heterogeneous and barbarous troops, drawn from nations all of whom were inferior to the Romans, moulded a victorious army, that the Romans themselves could not withstand.

But the case is different in civil affairs. The most profound political philosophers of Greece, Socrates, Plato, and Aristotle, lived at a period when their counsel was most needed by their country, and we should suppose would have had great weight, yet practically their opinions made no impression on the public mind ; their influence seemed to be limited to the disciples who attended them, and as respects even these, the result would seem to have been merely theoretical, if we may judge from the characters of Alcibiades and Critias, two of the most celebrated pupils of Socrates.

Aristotle, who analyzed all known political institutions with great acuteness of research, lived to see the extinction of all the free governments of Greece. Military force was, indeed, the instrument, but if the people had been imbued with sound principles, and a proper sense of what constitutes a good government, and the value of freedom, no power would have been sufficient to subvert the Grecian commonwealth.

A man of great prestige, of commanding presence or popular gifts, may have influence with the commonalty, but his sway is rather by impulse, and leaves no lasting impression, if there is not a sound basis in the character of the people.

Individual character is moulded by domestic training and associations of early life. Without such training evil propensities will gain the upper hand, and a disposition naturally good be **perverted** to the worst purposes. Then it is to be also noticed that character is hereditary, and that vicious habits will descend **to the** offspring of vicious parents, **if** not controlled **by** powerful influences. Education **may do much, but the most** important part of education **is example.** In a depraved community it is exceedingly difficult **to instil virtuous** principles into the youthful mind. The **nature of man is** prone to vice rather than virtue.

So with national character ; the people **is made** up of individuals ; and where there is no individual discipline, no domestic school of virtue, there will be an aggregation in the mass of all the corruptions of the individuals composing it. Socrates ironically said, that *to* **have** *respect for* **the** *opinion of the multitude was like rejecting a coin as spurious, and taking a large quantity of* **the** *identical* **coin as** *good money.*

We see at this day **in the French the want** of domestic training. There have been many learned **and** able men in France, but they **have** had little moral influence **over the** popular mind. There is in that **country a lack of** *individuality ;* few think **for themselves, and the** impulse by which men **are moved, who have only a** social not **an individual** character, **is** something that stirs **only the** most mobile feelings—it is **the** sensitive **and** superficial, not the **interior and** reflecting part **of** their **nature**

that is acted upon—*wit* has, therefore, **more motive** power than *profound thought.*

The Athenians, after they had become profligate in private life, could be roused to sudden efforts by their orators, and even become good soldiers under **a** distinguished **military** leader, but **could** not be reformed in their private habits by the instruction **of** Plato, or the example of Phocion. The commonwealth suffered from individual vice, **public** business was **conducted with levity, and the** attendance of the people **in the** legislative assemblies **or** the dikasteries, was rather for entertainment **than the** discharge of a grave duty.

Even with all the **aids of** education, it is needful **that there** should be respect for the eminent men of former times ; and there should be something in the recollections of the past deserving of admiration, which fathers may with pride rehearse to their children.

The love of country needs **to be** sustained by the memory of greatness and **virtue.** The blessing of freedom must be hallowed by the recollection of the perils incurred, and the achievements performed, in establishing and maintaining it.

3

VII.

LAWYERS.

It is a question, what is the moral influence of the indiscriminate defence of right and wrong by *hired pleaders?* It is a natural consequence, if a man hires out his services for the defence of others, that his zeal will be in proportion to the reward ; and if he may, for such a motive, put forth greater casuistry or pathos, it will be easy to justify to himself any contrivance for success. We have instances, in desperate cases, of deliberate devices of the most unscrupulous character, for the escape of the criminal. The lawyer, in such a case, is held to but slight account, though all others concerned would be considered guilty of a misdemeanor.

I know there are some who profess to discriminate, and to avoid unworthy clients, but I have known no instance of a large fee being refused on account of the character of the party. The effect on the public mind is bad, because it is a practical example of the perversion of truth, by men well educated, and having considerable influence in society, but more especially of the actual obstruction of justice by venal skill.

Cicero prescribes, as a moral rule, that the defence of all criminals may be undertaken, with the exception of those who had committed crimes against religion, or some great atrocity ; and that in the trial it is the office of the judge to seek for truth only, but that the advocate may be permitted to maintain a probability of truth even in the support of what is false.*

The orators of Athens and Rome were not strictly lawyers, while a popular government subsisted. There was, indeed, no occasion for any great amount of technical knowledge, as the courts made little account of precedents. Judicial magistrates were, to a certain extent, subject to uniformity in the rules of decision, but the Athenian *dikasteries*, which were composed of the people, had absolute power in each case, with no restraint in the nature of an appeal. The Roman Prætor was required to publish, on going into office, a declaration of the rules by which he would decide cases, and as he held office but for a year, this was annually renewed ; but he seems to have been at liberty to exercise a large discretion in the adoption of rules, though practically, as we may infer that the same set was continued, with only occasional modifications and additions. This was the origin of the Prætorian

---

* Judicis est semper in causis verum sequi—patroni *nonnum quam* veri simile etiam si minus sit verum defendere.—De Off., lib. 2, c. 14.

Edict, which was finally made **perpetual** in **the reign**
of Hadrian. It is obvious, therefore, that causes
could be tried at Rome with very little legal learning. In fact, a lawyer was not **one** who was versed
in judicial decisions, as these were not binding **on**
the courts, but one who was acquainted with the
forms of actions and the technical language of laws,
with the addition, in individual instances, of a
knowledge of all the laws which had been **enacted**
by the Senate or people, **which, in** English **phraseology,** would be called statute law.

When the orator had occasion for any professional
information of this kind, he applied to a lawyer **as**
he did to an architect, a physician, or geometer, for
whatever knowledge was needed in the respective
trades or sciences with which these were conversant.

Pompey, ***Cæsar,*** *Crassus, Lucullus,* and other
generals, prosecuted **and** defended **criminal** and
civil causes—which proves **that no** considerable
amount of professional knowledge was required, as
the men I have referred **to were** bred wholly in mili-
tary and political affairs. Under **the** imperial government the characters of orator and lawyer were
combined. Public speaking was then limited to the
courts, and **law, as** a profession, had made a great
advance in respectability. *Seneca, Tacitus,* and
Pliny, were lawyers. It **is a singular fact, that the**
science of jurisprudence **flourished in** the highest
degree under some of **the** worst emperors. The

classic age of Roman law was between the reigns of
Hadrian and Alexander Severus, a period that produced such monsters as Commodus and Caracalla,
but which is made illustrious by the most celebrated of the Roman lawyers, *Gaius, Ulpian, Papinian, Modestinus,* and others.

The lawyers of England have had an important
part in the civil history of that country. In general they have supported royal prerogative, but there
have been some notable exceptions. *Coke* was
prominent, as a member of Parliament, in resisting
the arbitrary pretensions of King James, and preparing the way for the subsequent popular reaction
against the measures of Charles First ; but the motive by which he was at first actuated was probably
resentment at the preference given to his rival,
Bacon.

Somers took an active part in the revolution of
1688, but, after the settlement of the succession,
the lawyers and judges have generally been sup-
porters of the extremest pretensions of royalty.
The reason for this is obvious, viz., that the ap-
pointments to office depended upon royal favor. A
few exceptions occur of lawyers who have been so
confident in their own abilities as to risk the dis-
pleasure of the crown, and who forced a reluctant
concession of preferment from the king. Among
these the most conspicuous are Dunning, Erskine
and Brougham.

American lawyers are a class of men differing much from the profession in any other country. The source of preferment in the United States is popular favor. Lawyers are almost universally politicians. Most public offices, administrative as well as judicial, are obtained by lawyers. There is certainly no want of ability. It is no more than just to say that the American lawyers have legal learning that will bear a favorable comparison with that of the bar in England, and that they have a readiness in the dispatch of business, and a versatility in combining other pursuits with the professional, which has never been equalled in any other country. To this praise there are, however, some serious drawbacks. Being generally politicians, and aspirants for office, or at least for political influence, the consequence is, they are unscrupulous in availing themselves of all the means of accomplishing their ambitious purposes, and too apt to descend to low intrigue and management.

VIII

HEREDITARY CHARACTER.

THE care with which the English trace their descent, is attributable to aristocratic pride. The Jewish custom was derived from peculiar religious policy, and, like circumcision, can be explained only by the tenacity with which they held to a principle of distinction, which applied not only to the nation but to tribes and families.

I have derived some ideas on the subject of hereditary character from facts recently developed, showing an analogy in the law of human organism to that of the lower order of animal constitution much greater than what has been heretofore supposed.

Whether it be that the soul acts *only* through the organs of sense (as Locke supposed when he denied that there was thought in sleep), and from which would be perhaps deducible that the *soul* itself is but a subtle or ethereal form of matter, one thing is certain, that a man's character (by which I mean to express a certain individuality in thought and action) is intimately connected with his physical organization. If this hypothesis be well founded, hereditary character is demonstrable, for nothing is clearer than a resemblance of physical structure, both external and interior, in parents and children.

Education and circumstances in life may restrain our natural propensities, or give direction, in one course of action or another, and thus apparently determine character, but these, after all, have little power against an organic bent or inclination of mind when there is great physical energy, by which I mean not bodily strength but strong impulse, whether that be derived from inherent proclivity of mind or sensual organization. In such case restraint is overborne, or, if submitted to for a time, it is only as to a coercive power—and the natural temper, when it has opportunity, is apt to break out with redoubled force. Nor is it merely sensual passion that thus acts ; all the faculties and powers of the mind seem to have a like, though perhaps not equal persistence of action, and I am much inclined to believe that they are in a similar manner dependent, at least for their activity, on some physical principle.

The transmission of bodily diseases from parents to children has been long observed—as consumption, scrofula, gout, insanity. It is properly the transmission, not of disease, but of a corporeal structure, which is liable to such diseases.

The external resemblance of children to their parents is a general law ; sometimes the resemblance is to ancestors more remote ; so it is in respect to *character*. Peculiar family traits **may be** traced through many generations.

The *Claudian* family of Rome **is a** conspicuous instance, which for many centuries was the most haughty and aristocratic of the Patricians, and finally became the **tyrants of Rome.** From Appius Claudius, the Decemvir, to the monster Caligula, the same imperious temper seemed to pervade the race, or at all events, to be often reproduced in **in-** dividual members of the family.* The Catos were, during several generations, equally remarkable for severity of rectitude, from Cato the Censor to his great-grandson **of** the same name, who killed himself at Utica, and Marcus Brutus, the nephew of the latter.

The Guises of France were, during at least three generations, alike in their imposing stature, **seduc-** tive manners, and factious disposition. The same traits descended through Mary of Guise to the celebrated Mary Stuart and her posterity.

The Stuart family of Scotland are known, historically, as having displayed a singular obstinacy or inaptness to yield to changing circumstances, and thereby **suffering** great misfortunes. Queen Mary lost her throne and life—her grandson, Charles First, of England, came to the same end—his son, James Second, was dethroned, and the family, after its exile, still continued intractable as before.

---

* Quoted from Gregory's Conspectus Medicinæ Theoreticæ, in " **Combe's Const. of Man,**" p. 146.

The transmission of a morbid temper of mind is illustrated in the poet Byron. The family, from the time it became historically known by the grant of Newstead Abbey to Sir John Byron, by Henry Eighth, had the characteristics of recklessness and extravagance. Charles I. granted a title of nobility and additional land, the family **having** before that time been **much** involved in pecuniary embarrassment. The grandfather of **the** poet, Admiral Byron, was brave but unfortunate—his great-uncle **and predecessor** in the title and ownership of the estate, killed his neighbor and relative, Mr. Chaworth, in a duel, and, as was alleged, by unfair means; ill-treated his wife, so that she was obliged to separate from him—wasted **his** estate, and lived solitary and friendless—always **went armed,** and supplied the place of his **wife by** a female domestic, who had the soubriquet in the neighborhood of " Lady Betty." Captain Byron, the father of the poet, ran away **with the** wife of the Marquis of Caermarthen, before he was of age ; after her death **he** married Catharine Gordon, the mother of Lord Byron—squandered her property, and by bad treatment forced her to live separate from him.

These ancestral traits descended to the poet, intermingled with the passionate temper of his mother. How he could have become possessed of *any good quality* seems strange, as his mother seemed to be endowed with little or none, **and his** father **was a** sensual, selfish, **and** unprincipled man. But **the** transmission of character by hereditary descent sometimes overleaps one or more generations. **He** had the solitariness, gloom, and domestic irregularity of his great-uncle, and he may have derived his better qualities from **a** source more remote.

Voltaire mentions a case, within his own knowledge, of **a** father and two sons each committing suicide at the same age, and without any known cause.*

Dr. **Burrows** relates a family **trait of the same** kind exhibited in three generations—the grandfather hung himself, three of his sons destroyed themselves, **two of** the grandchildren followed the example, and **the** fourth generation showed symptoms of the same propensity.†

The mother seems to have most influence on the character of the children. If she is weak in mind, *the offspring will exhibit a deficiency of intellect, even if the father have **more** than ordinary vigor. **On the other** hand, if the mother have great energy and any peculiar traits, they will be reproduced in **some degree in the children,** even if the father be of ordinary **or** inferior character.

It **is almost** proverbial that a distinguished **man** is always found to have had a mother more than ordinarily endowed with vigor **of mind.** The care of

---

* Phi. Dic., article "Cato." † " Burrows on Insanity."

a child in its early years is indeed of much consequence ; but if the mother have good qualities she will impart them to her offspring at their birth ; these will be fostered by maternal discipline, but will be seen to some extent, even under the most adverse circumstances, as the premature death or physical disability of the mother.

The Gracchi, the Emperor Constantine, Charlemagne and Napoleon, are familiar instances of greatness which seemed to be derived chiefly from the mother. The innkeeper's daughter, Helena, mother of Constantine, was indeed of humble origin, but the veneration which the emperor always exhibited toward her, even in her old age, is a sufficient proof of her remarkable qualities.

Edward Third, of England, derived from his mother, Isabella, his gallant and enterprising character, although she was not a pattern of domestic virtue, but he inherited also her amative propensities.

The warlike sons of the Duke of York (Edward Fourth and Richard Third) must have owed their energy to their mother, who was an extraordinary woman.*

On the other hand, the degeneracy of the children of distinguished men, which is also proverbial, may be derived from the weakness or vice of the mother.

---

* I recollect a very interesting narrative of her life, in a fiction called " Cecilia, or the Rose of Raby "—from what source the materials were derived I do not know.

The wife of Socrates brought him a stupid family—the wife of Marcus Aurelius produced a circus rider, instead of a philosopher, but it has been a question whether the Empress Faustina did not find a parentage for Commodus in one of the performers at the amphitheatre.

I know myself the family of a former eminent judge of the Supreme Court, U. S., who, to the second generation, have been all weak in intellect, and it needed only acquaintance with the mother (who was of an aristocratic lineage) to account for it.

There was some sense, according to this principle, in the proposition of Hortensius (as related by Plutarch), of getting the loan of the wife of Cato, or his daughter (who was also married). By such intercourse with women of virtuous families, he said, there would be insured a virtuous offspring.

As to physical advantages to be considered in reference to marriage, Sir Walter Raleigh advises well :

" Have a care thou dost not marry an unseemly person, for comeliness in children is riches, if nothing else be left them ; and if thou have a care of the races of horses and other beasts, value the shape and comeliness of thy children before alliances or riches."

IX

SENSUALITY.

THERE is some plausibility in the doctrine of the Shakers, that the sin which occasioned the fall was sexual—certainly there is no propensity more universal or engrossing.

The lower class seeks sensual gratification for itself; the more educated and well-bred observe some forms of delicacy, and usually veil the natural desire under a certain "illecebrae"—the playfulness of love—the familiarities of friendship.

But whether a hirsute appetite or refined sensualism be the phase, there is in unchecked indulgence a destructiveness to bodily and mental vigor. The former is exhausting to corporeal elasticity—tends to intemperance, and induces a coarse, selfish tone of mind, that is insensible to a refined voluptuousness. The other engrosses the imagination and titillates the senses by pruriency of thought. A late bishop of the southern diocese of N. Y. may serve as an example of a whole class; such men find a sufficient exercise of meretricious thought in immodest familiarities; but there is in such a habit a depraving influence, fatal to a healthful condition of mind.

Others there are who indulge in a morbid sensualism of thought, without actual **licentiousness** in conduct, of which class of **men the historian** *Gibbon* is a prominent instance ; **even** in a man of *erotic* **temper, rigid restraint may be** attended **with de**rangement **of the whole organism.** Many pious **men,** in struggling against passion **which** they deemed sinful, have been involved in insanity. Luther's opinion as to celibacy is well known ; it must, however, be considered **extravagant.** There are men who, without lubricity of mind, have **lived chastely** in a single life. In such instances there is **a well**balanced mind and **a** happy physical organization ; **by** which I mean **one** of delicate structure—one sufficiently strong **to** bear the ordinary roughnesses of **life,** but without the coarseness and solidity necessary for the endurance of great hardships. A genial, philanthropic, sympathetic temper, combined with intellectual and moral endowments, **is seen in** such a one **as I have described.**

If indifference arise from physical defect, or if there be an aversion growing out of ill-reception, by reason **of** deformity, or other like cause, there is usually exhibited a malevolent spirit. The eunuch is an ill-natured being. The biting sarcasm of Swift may, perhaps, be attributed to "a tentiginous humor repelled to the brain," as **he** expresses it.

Sensual passion may be controlled by other engrossing **objects** ; but generally great mental appli-

cation must have an equivalent, or physical enjoyment of some sort. The more abstracted the mind
is, the more intensely it is fixed upon the object of
its thought, the greater will be the compensation
required in some grosser element.
—Great thinkers, it has been said, are great sensu-
alists ; but this must be taken in a qualified sense.
There may be a dreamy sort of abstraction which
involves no effort ; it is rather the absence of
thought—the indulgence of passive sensation and
semi-sensual ideas flowing without restraint or method. But if the mind be disciplined to graver and
more toilsome employment—if it be fixed upon some
specific subject, and tasked by consecutive and me-
thodical reflection, there is an exhaustion of the
material organism, or some part of it, perhaps of the
finer part of the nervous system, which by consequence requires rest. But rest is obtained by alternation, that is, by the exercise of other bodily functions.

Sleep is the nearest to absolute rest, but even *that*
is not perfect ; the vascular action is in that state
increased, and the organs of absorption, secretion,
and assimilation, are more vigorous in their action
than when we are awake. Yet sleep is a rest of all
the faculties of body and mind which are subject to
voluntary effort. The partial rest which results
from an alternation of employment is most effective
according to the change or contrariety ; therefore, the

tension of the nervous system by thought demands
muscular exercise and sensual pleasures.   Hence,
excessive exertion of mind ought to be avoided. Religious contemplation, if not fanatical, has a tendency to promote serenity, and therefore demands
no such compensation ; and this is the great excellence of a consistent religious life. But inordinate
zeal and fervor, that cannot be kept up as a general
habit, must, by the law of our nature, be followed
by a reaction to the other extreme.   Hence the con-
trarieties in the lives of many Christians, which they
mourn over as the necessary consequences of sin, but
which are in fact in a great degree attributable to
their own indiscretions.

A biographer of Sir Humphrey Davey refers to
the sensual habits of his subject, but which were
limited to the comparatively innocent practices of
epicurism in eating and a fondness for fishing.   A
critic (in the *North British Review*) makes some
sensible remarks, which I transcribe : " Almost
every great man is a voluptuary by nature. Even
Newton smoked himself into a state of *etiolation*.
Your true consumers of tobacco, your genuine gour-
mands, your consummate lovers of wine, your most
absolute of gallants, and your only sufferable opium-
eaters, are such men of genius as do mostly toil like
heroes when they are at work.   Doubtless men of
genius are endowed with the most sensitive and
quivering of corporeal frames, and if their charac-

ters be at the same time strong and vigorous, that swiftly responsive constitution to the play of every sensuous delight is invariably accompanied by the fiercest manifestation of turbulent human passion. * * The mind which is overstrained instinctively seeks and finds its natural repose in the pleasures of sensation, and the wearied sense aspires to hide itself in the kindlier bosom of emotion, whence the intellect springs anew in renovated strength."

Lord Stowell (an eminent English judge, brother of Lord Eldon) was a great eater, and would drink two bottles of port at one time, and seemed not to be injured in health by such habits. He was curious upon all subjects, and the most indefatigable sight-seer in London. Whatever show could be seen for a shilling or less was visited by him, and he was often seen, after his elevation to the bench, coming out of the penny show-rooms in the streets of London.

X

HEALTH.

THERE is an art of life, and he who understands the inner principle upon which his condition depends, and not merely the external development, will find a beauty in human life which is worthy of its great author, and the material out of which happiness may be wrought for himself.

Health is the first, and an indispensable constituent ; not the health of the day-laborer, who can work twelve hours without exhaustion, for that is strength of sinews only, which the ox has, and perhaps with as great a sense of enjoyment, or if there be a difference, it is that the human animal is prone to indulge in harmful gratifications, as in strong drink and other sensualities. Nor is it the health of one whose stomach can bear tasking much beyond its proper office, and counts a dinner a chief enjoyment of life, first, in the anticipation of the pleasure and then in the slow and protracted fruition, for this is the existence of an oyster, or of a boa constrictor ; " total extinction of the enlightened soul."

The health that I refer to is a condition of mind and body at all times or habitually susceptible of agreeable impressions, and therefore requiring sensibility or delicacy of external organism, and of the interior nervous structure—cultivation or discipline of the organs of sense and of the faculties of mind, so as to avoid the extremes of mere physical sensation, or of over-refined emotional susceptibility—and so as to be furnished with pleasing impressions from all external objects, and an equally pleasing consciousness from the exercise of thought.   Harmony of all the faculties of mind and body, when those faculties are educated, is the true state of happiness.   Disease impairs enjoyment, at least of the placid or habitual kind which is most consistent with long life, but may, by rousing into more than usual activity certain powers of mind or body, whether by great facility in their exercise, or by counteraction to the disturbing force, give to them more acute sensibility, and by consequence, more exquisite but less continuous pleasure.

Another requisite is, that this sensibility of nerves should be natural, not morbid.   A bodily constitution that is " servile to every skyey influence," and suffers a shock from even ordinary incidents of life, is predetermined to the extremest human misery, often ending in the unuttered woes of madness.

I have observed, in many instances, an alternation in the course of disease between the mind and body ; sometimes a deranged function of the one is relieved by transfer to the other.   Perhaps, when the mind becomes thus affected **by** alternation, the nervous system is the seat of the transferred disease. This is the part of the material organism which is nearest in contact with the **mind.** There may, indeed, intervene some subtler material, not discernible **by** our senses, **as is** recently held **in** respect to light, that there is **a** more rarefied medium than the **air** whose undulations produce the effect which we call vision. One phenomenon has been noticed in respect to this alternation of disease **between mind** and body, that sometimes when **the latter has been** relieved of **a** chronic ailment, a change **of character** has taken place ; **evil passions and propensities have been** developed **where before there had** been a mild **and** amiable temper **of mind.** *Hahneman* says that the **new** development is in fact the revival of qualities previously existing, but which had been kept down by bodily disease.* **It** is, however, **a more** natural hypothesis to consider the mind as now the seat **of** the disease which **had** before affected **the** body.

Plutarch relates an anecdote of a soldier of Antigonus, **remarkable for** bravery, but **who** had an unhealthy appearance. **On** account of his courage, Antigonus put him in charge of his own physician, who succeeded in curing the disease ; but the character **acter** of the soldier became entirely changed, and he

---

* Organon, p. 173.

no longer exhibited his former bravery. This being observed, and the reason asked, he said that he was made less bold by being relieved from misery, by which his life was made hateful to him.* The story may be fabulous, but something analagous has doubtless fallen under the observation of every one. The bold and adventurous are generally those who have suffering of mind or body. Our thoughts revert to *Wolfe*, wasted and almost dying with dysentery at the time of his desperate but successful attack upon Quebec—to *Nelson*, at Trafalgar, mutilated in former encounters with the enemy, having but a single arm and eye, and now seeking only for a glorious death.

It may be that the outward sense, being reduced by bodily infirmity or great suffering of mind, which oppresses the functions of the body, dispels timidity, yet in a low state of the nervous system we often see womanish fear. We cannot wholly analyze the secret operations of disease ; some men, no doubt, are made fierce by derangement of the bodily organism ; the natural temper of the mind, we should suppose, would have much to do with such a development ; but as we see an insane man fearless of consequences, so may nervous derangements, operating upon some faculties, give to them a morbid perversity and an unnatural energy.

---

* " Life of Pelopidas."

How essential a sound bodily condition is to success in the world, indeed to the development of all great qualities, though somewhat trite, it may be interesting to consider in some particulars. "To be weak is miserable doing or suffering," although represented by the poet as spoken by Satan, is an axiom applicable to human life. The strength which is needed is not, however, necessarily of bones and sinews, but the physical vigor which is required for putting into practical use the powers of the mind. A public speaker must have not only strength of voice, but nervous energy sufficient to bear the waste induced by the effort of body and mind, so as to avoid the disagreeable exhibition of exhausted vitality—the eye without speculation—the face without expression of thought. In conversation something of the same endurance is requisite to sustain the vivacity which social impulse calls forth. In order to have influence on another mind, quickness and energy are essential qualities; persuasion does not depend so much on sound argument as upon adaptedness of what is said to the particular occasion—quickness of application and facility of expression; exuberance of spirit has of itself a momentum; disjointed altogether from intellectual power, it becomes a mere physical force; it may intimidate but not persuade; but with the adjunct of some vigor of mind, even though moderate, it will wield a greater control than the more logical thought of a nerveless man.

An author is physically tasked by the labor of composition ; intense application of mind is in itself exhausting. No one can have success as a writer who has not bodily vigor adequate, not merely to sustain the labor of thinking, so as to preserve equanimity, but also to bear the exhaustion caused by the mechanical work.

In fine, sound thinking depends much on sound health ; irritability of stomach, or any disease affecting the temper of the mind, must impair clearness of thought, and the social sympathies which give practical direction to thought. There is, indeed, what may be called a physical strength of mind, which has not much to do with a clear perception of right and a conscientious regard for it ; or if there be clearness of view, it is that which comes from entire absence of reflection, so that there is no occasion for doubt. This phase of mind is singularly unimpressible by argument, and whether it be owing to obesity of understanding, or an austere rigidity of will, is a low order of character, nearly allied to the animal, and is generally found conjoined with a coarse, unpliable, muscular frame.

Then again there is the strong will, fixed by some intense master-passion, and this may be seen in a feeble body ; often, indeed, it may be that the body is too weak for the effort it is put to by the energy of the mind. In general, strength of character has to be sustained by some habitual resource—a recurrence of the mind to **certain** modes of thought peculiar to **the** individual, either **by** nature so **strong** as originally to control **his** mental processes, or **which** have become familiar by **long use.** As Wisdom **is** represented by the **poet as seeking** " retired solitude, **where she** plumes **her** feathers **and** lets grow her wings," **so the** most active mind finds occasion often **to** withdraw from the collision of **the world to its** own secret thoughts **and** impulses, **from** which **it** derives its aliment. **By these its self-complacency,** which is continually **impaired by the rough passages** of life, **is again restored, its cherished** purposes **are** renewed, **and a new energy imparted.** Various are the resources **of** different **men,** but whatever they **may** be, the resort to the habitual source of strength **is like** that of Antæus to his mother earth, as represented in the ancient fable.

Capacity to bear the fatigue of mental labor **depends,** as before remarked, upon nervous **energy ;** so also the power of enduring **pain of body or** anguish of mind. Those, however, that **we** call nervous people, **have not** strong nerves in **the** sense now spoken of ; **they have** sensitive or irritable nerves, without power of bearing any unusual action, or of resisting or controlling **any** disturbing force. Ill health may weaken **the** nerves, and this is generally **so,** though there may **be** disease not affecting vital organs which will not impair vigor of mind. Diseases **of** the liver are most fatal to mental energy.

4

.

Under the theory which I have suggested, it is evident that incessant strain of mental powers, that is to any greater effort than is natural, impairs their strength, as hardship and overstraining the **bodily** system will impair the strength of the latter. Overtasking either mind or body for a present object is, therefore, in a prophylactic **view,** short-sighted policy. Better is it to forego for a time, **or** even altogether, some part of what we aim **at, than by forced** effort waste our vigor without the possibility of **re-**paration. Impatience to accomplish immediately some object of desire, whether it be the acquiring of a great name in the world, the accumulation of **a** fortune, or whatever else may be intensely sought, **is** the origin of **much** irregular and misappropriated action. This evil is observable in the habits of professional men, especially the clergy of this country, whose course **of** study **has** little relief by variety. A clergyman is expected to give himself up entirely to the duties of his charge, and has consequently little leisure for rest of the mind, **or** even for doing deliberately what requires thought and pains-taking.

A paragraph from an article in an English Review upon this subject has much force :*

" In this respect, **to** be sure, the fate that has overtaken the clergy is only the same that has fallen upon every order of men—upon the medical profes-

---

* *Frazer's Magazine :* 1847. " Recollections of Dr. Chalmers."

sion—upon the profession of law—upon ministers
of State—upon members of the legislature—all
over-worked, driven on as by the force of a hurri-
cane, which leaves them no faculty of deliberate
thought, not even the time to go through the business on hand, except in the most perfunctionary and
inefficient manner. It is a miserable system, which
must ere long, unless it be checked, prove fatal to
the best interests of the country."

There is in the progressive change of the human
organism, from youth onward in life, certainty in
the midst of uncertainty. Nothing is more certain
than that within a period of four score years, or, in
a few exceptional cases, a little upward, the infirmities of old age will supervene ; but nothing is
more uncertain in respect to any individual than
what point he will reach within this ultimate limit.
Accidental causes may bring on premature decrepitude that naturally belongs to advanced years. The
vigor of youth may suddenly fail under the blight-
ing influence of vice, or of latent disease descended
from ancestral vice or improvidence. Any rule for
the regulation of life must have reference to the exception as well as the general course of nature.
Disappointed hope, reverses in life, or whatever disturbs or oppresses the mind, undermines also health
of body, and few there are who do not meet with
some of these before the set time of natural decay.
Irrespective of such disturbing forces, there is in

every human being a certain degree of vital energy, which will, in regular course, last its specific time, and then run out. We indeed imagine, in early life, that there is in us a self-renewing power that promises perpetuity, and so there is to a certain extent at that period, it being provided for the growth and perfection of the body ; but in middle life we become sensible of the want of **the** recuperative energy which had before enabled us to resist many rude shocks, and to rally from a state of prostration ; further on in life the vital power, although adequate to the ordinary wear of the corporeal mechanism, becomes more feeble in its renovating action. Disease or bodily injury of any kind is then more dangerous ; the sorrows of life have a more tenacious grasp ; they are, indeed, resisted with more apparent firmness than belongs to youth ; but the resistance, though it keeps up the semblance of strength, often ends in a sudden and fatal issue.

From this constitution of our nature it is in early life sensuous, because overflowing with surplus vitality, and the mind partakes of the same exuberance. Our thoughts and feelings are then associated with **objects** of sense, and we can scarcely conceive of a state of enjoyment independent of the sensual element. Afterwards we learn that what was so enchanting to our youthful imagination is a source **of pain as** well as pleasure ; and, if our experience be under proper direction, we are compelled to seek for happiness more in inward and permanent consciousness than in the external and changing. Thus is there a process of discipline which, wisely regulated, moulds the character of every man into distinctness and almost isolation. Fortunate will it be if a wholly selfish convergence of thought shall be avoided, while the mind finds in itself its power of sustentation and chief sources of enjoyment, yet has sympathy with all other natures congenial to its own, and a sense of pleasure reduplicated by being imparted to or enjoyed with them.

Then follows infirmity of body, which gradually alienates the mind from its corporeal association, **and** leads it to contemplate a spiritual state of existence as the only permanent condition. Intimacy with the few early friends that may still survive, will maintain its hold and become, indeed, more sacred than ever before ; but this is all that remains of earthly affinity. Thought of the future, and of the state of the larger number who have gone before to a mysterious, unrevealed condition **of** life, and a stronger reality of the relation of the soul to its divine author, are more and more present to the mind with advancing years. Weakness of the body, whether the effect of old age or prematurely induced in earlier years, is a powerful preparative for religious thought. The healthful, especially in early life, feel self-sufficiency in their strength ; the old and infirm feel a sense of weakness, and naturally **look to a support out of themselves.**

It would be interesting to consider more in detail the influence of weak health in the formation of a religious character ; for the present I shall, however, only remark that not merely is a religious turn of mind promoted by bodily infirmity, but some peculiar phases of opinion and feeling are generated thereby, which are perhaps erroneously attributed to a mistake of judgment.

As religion has to do with feeling as well as opinion, it cannot well be otherwise than that its development should be largely influenced by the state of health. I cannot help thinking that the best form of piety is not that which has been wholly brought out under bodily suffering. A spiritual frame of mind is indeed the highest source of consolation in suffering, but there can hardly fail to be some perversion, by the sympathy which the mind has with the body. Even the well-tried Christian, who has, in the vigor of life and the full maturity of his powers, been devoted to the service of God, will, under the influence of disease, sometimes fall into despondency. His faith will, indeed, not fail him, nay, perhaps may be even brighter at the last ; but, in protracted illness, human weaknesses will, to some extent, be intermingled with and give color to divine truth. Then, as to the enjoyment of religious hope, there may be sometimes seen in the midst of suffering a cheerful and even triumphant elevation of mind, like that described by the apostle

Paul "as sorrowful, yet always rejoicing, ✿ ✿ as having nothing, yet possessing all things." (II Cor., 6–10.) **Yet** this **is a** high order **of faith, and** may require, perhaps, a certain degree **of physical** strength and firmness. **There are,** however, many other cases where there is a low state of feeling—a self-distrust, and want of a clear and comfortable **hope.**

Some nervous **diseases** overcast **the** mind **and give** a sombre aspect to all the thoughts. **If it may be** so in respect **to the** present life, **is it** strange **that it** should be the **same** with **what lies beyond, of which we** have **a less vivid sense?** Dr. Moore has observed **that "there is many a** fine spirit **so** mistaken as to gather clouds about its faith, which obscures the light of heaven, and whose conscientiousness causes the feelings of the body, opposing and distracting the better desires of the mind, to seem like the **wit-**ness **in** themselves **of a** perpetual condemnation ;" and he suggests, for the consolation of **such** persons, that there are impressions upon **the** nervous organ-ization which inevitably affect the mind, but which are sinful **or** otherwise just **in** proportion as they are indulged or resisted.✿

As to regimen **of life** for the conservation **of** health, much has been written, **but I have found** little advantage from all **I** have read, except as sug-

---

✿ " Use of the Body in Relation to the Mind."

gestions in aid of **my own** observation. **Some** general axioms there are which have application to all men, but for the most part what is chiefly available is that which a man has himself observed in his own experience, if he has the habit of attending to what affects his well-being.   After all, it is not so much want of knowledge as **want** of firmness to pursue **the course** which is known **to be best,** that is the **cause of** most evil practices **in the world.**

Among the directions **relating** to health, **I think** those are of chief consequence which have in **view** the mutual influence of mind and body.   Health, according to the definition I have before given, involves both mental and corporeal soundness.   As to **habit of mind,** interchange and variety suit best in **youth ; uniformity in** advanced **life.**   In fact, habit becomes **of itself a** positive **enjoyment** unless vicious, and it is well **if all a** man's **habits have** been properly formed, for **there** is the **same** tenacity of the evil **as** the good, although **not equally** the source of **pleasure ;** on the contrary, the **former** are the **cause of** misery, **yet** seldom given up.   Bacon recommended, **" to entertain** hopes, mirth rather than **joy, variety of delights** rather than surfeit of them, won**der and admiration, and** therefore novelties ; studies **that fill the mind with** splendid and illustrious objects, **as** histories, fables, and contemplations of nature."*

---

* Bacon's Essay, " Regimen of Health."

As to discipline of the body, there are certain commonplace rules which apply in a state of health, but are unsafe if there be weakness or disease. *Cold bathing* gives vigor to a constitution which is naturally healthy, and may be of service even where there is weakness, if there is no organic disease. *Exercise* is not equally of use to all. A quiet condition suits better with some persons, particularly where there is general weakness. Diet, in many cases, is more important. Much exercise does not appear to conduce to long life. A habit of ease, so that no hurtful indulgence, as the use of strong drinks, &c., be added, on the whole, is most favorable to longevity.

4*

XI

NARCOTIC STIMULANTS.

TOBACCO, OPIUM, **INDIAN HEMP, ETHER,** CHLORO-
FORM, &c.

IT is uncertain what acquaintance the ancients
had with the narcotic principle. *Hellebore* is spoken of by Greek and Roman writers as producing
madness, and also a cure for maniacal diseases ;
which use of it must therefore have been in accordance with the modern homœopathic theory. It
seems also to have been taken sometimes for a supposed property of giving clearness to the mind.
Mandragora, commonly understood to be the plant
now known as the mandrake, but which hypothesis
is not sustained by satisfactory proof, was the subject of superstition which has extended even to later
times. It was supposed to have miraculous powers
and to be be used in witchcraft—a fabulous property was also attributed to it of producing sensual
fertility.

The most powerful narcotics now used have been
introduced within a recent period, unless there may
be some truth in the conjecture that the *hemp* was
known by the priests in Asiatic nations, as it is certain that some stimulant, powerfully affecting the

mind, was resorted to, particularly in Phrygia, **the knowledge** of which may have been carried **thence** to Greece and Italy.

A classification may be made into three distinct orders of stimulants : 1. Those which have **a sedative** effect of so mild **a** character as to admit of habitual use, and though in the end resulting in injury, yet not immediately destructive to the vital functions, **of** which class *tobacco* **is** chiefest **in im**portance.

2. Such stimulants as have a nervous **influ**ence, when **used in** any quantity, if it be **done** habitually, of **which** *opium* is the principal. It **is used largely in** the East, and **to a** considerable extent **in Europe, and** when once the habit is contracted **it** is rarely overcome. Although the first effect is a pleasing illusion and a seeming aerial action of mind, yet the ultimate result is weakness **of** all the intellectual powers. Among the Orientals, especially the lower class, **a disgusting** stolidity, often even idious, may be **seen.** We have some memorable instances of the **use of** opium in England, which would seem to show that intellectual vigor may be maintained, despite the baneful consequences I have supposed. Yet it may be observed of Coleridge and **De** Quinsey, that their efforts **were** spasmodic — that both **of them** had **a** dreamy **va**cancy of thought at times, whereby considerable part **of** their lives was wasted ; and again, **it** can

hardly be estimated how much temporary resistance to the indulgence may have been enforced under great pressure—in other words, how much they may have been saved by the incessantly recurring necessity of restraint, in order to accomplish their intellectual tasks.

Hemp, now commonly known by the Arabian term *Hashish*, belongs also to **this** class, though it is still less susceptible than *opium* of daily use.   It has, indeed, **little** adaptedness to the civilized life of Europe and this country. The effect is too excessive in degree to admit of frequent repetition. According to the reports we have, the action of **mind** produced by it is wild, often amounting to **frenzy**—**add to** which there is always danger to a sensitive nervous organism.

3. Narcotics, whose direct tendency is to produce suspense of consciousness, and if taken in any considerable quantity, will be fatal to life—hence, in medical science, called *anæsthetics*, which signifies literally substances that destroy corporeal sense.

Opium might be included under this head, **as when** taken **in** sufficient quantity it produces coma and even death ; yet, used in a moderate degree, it has other effects, before referred to, which distinguish it from narcotics which merely diminish sensibility.

The other principal anæsthetics are *sulphuric ether and chloroform* — the former manufactured

from alcohol and sulphuric acid, the use of which
for alleviating pain was brought into public notice
in the early part of the last century, but, probably
owing to some imperfection in the mode of preparing it, did not come into much use till within a recent period ; the latter, obtained from the distillation
of chloride of lime and alcohol, was first introduced
into use in 1846, at Boston, for relief from pain in
surgical operations, and has since been brought into
general use in all cases where ether had been resorted to.   Chloroform was for a time preferred to
ether, on account of its superior power over the
nerves, but numerous deaths having ensued from its
effect, it is deemed safer to use it in combination
with ether.

It is not my purpose to discuss the medicinal
properties of narcotics, but to consider the effect of
habitual use. Of *opium* I know little, except from
what has been published and is familiar to most
readers of newspapers and magazines.

Of *tobacco* I have some personal knowledge, and
I can render no greater service than to notice some
of the prominent incidents of the use of this narcotic.

Example is better than precept. The narrative
of the close of a drunkard's life, when his health
has been shattered, his mind impaired, and the
horrors of penury, imbecility, alienation of friends,
and desolation of his family, have gathered around

their helpless victim, have done more to arrest the
first step of inebriation than all the moral persuasion
that could be brought to bear upon him.   The
" Confessions of an Opium Eater" startled the
fashionable world in England, by the vision of the
infernal conceptions resulting from the use of a per-
nicious drug, the first effects of which are singularly
pleasing to the senses.   It has been objected that
the writer of these " Confessions" has depicted too
attractively the delights arising from the first use of
opium, which beguiles its votaries into a continued
use.   But it must be a superficial mind that can be
allured by the insane joy attending first indulgence,
when, at the same time, are brought into view the re-
lentless craving and unavailing efforts to resist it
—the horrors which terrify the mind, and yet are
insufficient to rouse up resolution enough to withstand the iron grasp which drags the poor victim to
the renewal of scenes at which his soul revolts.

The confessions of a tobacco smoker or chewer do
not involve such strong lights and shades.   There
is no trance or spiritual vision to tempt the novice.
Indeed, to outward observation, nothing would seem
more opposite.   All the association is disgusting—
foul secretion—the unclean mouth—the vile spittoon
—no household gods could have been found for such
unseemly things.   The *Venus Cloacina* of the Romans—the divinity of the sewers, the great recep-
tacles of the filth of the city, would alone have been
appropriate.

Yet it would be a mistake to suppose that the use of tobacco, which to the outward sense is so revolting, has nothing attractive. There is a feeling of quiescence resulting from the narcotic power of tobacco upon the nerves.

Smoking, which is the most common form of the use of tobacco, is probably the most hurtful though not the most offensive. King James' *Counterblast* was not extravagant in designating it as " *a custom loathsome to the eyes, hateful to the nose, harmful to the brain, dangerous to the lungs, and in the black and stinking fumes thereof, nearest resembling the horrible Stygian smoke of the pit that is bottomless.*"

The first inhalation is not, indeed, nauseous, which may be attributed to a certain aroma interfused in the manufacture of the tobacco, but the ultimate odor is utterly repugnant to delicate sensibility. There is, in fact, something intolerable in the stench appurtenant to a room where there is habitual smoking—hence, in the hotels a room is specially assigned for this indulgence, and latterly in railroad travelling a car is given up for the use of those who cannot endure even a few hours privation of their habitual stimulus. My attention has been attracted, in my journeyings, to the many vacant seats of those who have gone to the smoking car, not a few of whom have had ladies in their charge, but have preferred the fumes of the segar to the social pleasure of their companions.

It may be worth our noticing the direct effect of different narcotics. *Opium* acts specifically upon the brain — *nightshade* (the *bella donna* of the Materia Medica) produces congestion, similar to what takes place by a ligature round the neck, preventing a return of venous blood from the head.

Aconite acts upon the sympathetic nerves, producing intense sensibility, wakefulness, anxiety, followed afterward, as described by a patient, by great clearness of memory and vivid imagination.

Tobacco affects the nervous system generally— more particularly the nerves of the stomach and abdomen. .

It is said by physicians, that all those substances which narcotize the nerves have more carbon than hydrogen—they seem to hinder the proper defecation of the blood by the air in the lungs. This observation will be found important in determining the relative injury of *smoking* and *chewing.* By the former the *lungs* are more affected, by the latter the *stomach.* It must be recollected that it is not a direct organic lesion of the lungs, but an interference with their proper office of purifying the blood, an important part of which process, it is well known, is the carrying off of the carbon of the blood by the air brought into contact with the lungs—the oxygen of the air uniting with the carbon

of the blood, and passing off in the form of car-
bonic gas. This process, it is obvious, must be impeded by the intermixture of the smoke of *tobacco*.
To some extent this effect must be produced by being in a room where tobacco smoke is inhaled by
another, and this of itself shows that the practice
is discrepant with the common **rules of** social cour-
tesy, which forbid everything offensive to **a** delicate
taste, much more what is positively injurious **to**
health.

The imperfect action of the lungs upon **the venous**
blood **is** indicated **in the** habitual **smoker by a**
blanched complexion, so singular as to have obtained
the name **of** *etiolation*—**which term was** applied to
Sir Isaac Newton, as descriptive **of a** peculiar appearance which could not be otherwise described,
and I have little doubt that the aberration of mind
which the biographies of that celebrated man admit
to have temporarily existed, without furnishing **any**
satisfactory solution of the cause, **was attributable to**
the fumes of tobacco, rather **than to the** overtasking of **his** brain by study.

The effect of tobacco, when **taken into** the
stomach, **which is** more excessive in *chewing* than
smoking, **but** to some extent takes place **in the lat-**
ter as well as the former, **should** also be noticed.
Persons having a slow digestion, and who may be
said to have a melancholy temperament, have, I
think, **a** predisposition to the use of tobacco, and,

having once commenced the use of it, rarely lay it aside.

Tobacco powerfully affects the stomach and vis- cera, and is sometimes used as a medicine for obsti- nate constipation. The effect when first used is excitant, followed by vertigo of the head and revul- sion of the stomach. By continued use the nerves lose their natural sensibility, **and** the gratification thereafter consists mainly in allaying an uneasy state of the nerves, caused by **the want of an accus-** tomed stimulus. When the natural constitution **is** not vigorous, loss of appetite will soon follow, and digestion, which at first seemed to be aided, will be impaired.

The effect of tobacco upon **a** healthy stomach, unvitiated **by** stimulus of any kind, is a distressing nausea. **Can** it be that a feeble organism does not suffer by it, although it may have become so insensi- ble as not to indicate it by the symptoms which are so readily developed in a healthy system.

It is a fatal circumstance that the more insensi- **ble the** nerves become the greater must be the quan- **tity** of the accustomed stimulant, and though no **more** than the **usual** excitement of the nervous system is produced by the increased quantity, yet the chemical action upon the fluids necessary for digestion must be greater. The secretions of the mouth suffer first by constant spitting, then by mix- ture with the smoke or juice of the tobacco. Saliva, it is known, is impoverished or rendered less fit for its office by excessive secretion. For the purpose of deglutition merely, it is supposed that any fluid might answer, but that the saliva is an important agent in the stomach. It is said to be proved by chemical tests that there is no difference between saliva and the gastric juice.

The muscular or membranous action of the stomach is important, and whether it be aided by the gastric fluids, as the bowels are by the pancreatic secretions, it is certain that it must be impaired by anything that diminishes nervous sensibility. It may be proper here to explain a seeming inconsistency. I have said that the sensibility of the nerves was impaired, and yet have mentioned irritability of the nerves as an effect of tobacco. The explanation is this : *A noxious stimulant first depresses the natural healthy action—this is followed by a morbid reaction.* The irritability of disease takes the place of the natural sensibility of health. *A schirrhous tumor may arise from imperfect function of a gland and remain long inert, and without sensibility, but in time the excruciating pain of cancerous action supervenes.*

It would be out of place to pursue these professional details further. They belong rather to a medical treatise. My object is to present, in a popular manner, some considerations which should satisfy a sound thinking man as to his duty in respect to the practice in question.

The magnitude of the evil is becoming developed in the alarming increase of nervous disorders. Let every parent of a family remember that though he may be willing, for the solace which his morbid taste finds in tobacco, to bear all the ailments it may induce upon himself, and think the evil more than counterbalanced by the enjoyment, yet will he entail upon his children a fearful preponderance of evil. *Nervous* susceptibility will be preternaturally developed in them, and it will be the premonition of a premature death or of life-long sufferings, and so is verified the old saying, quoted by the prophet, *" The fathers have eaten sour grapes and the children's teeth are set on edge."* (Ezek. xviii., 2.)

The same reasoning applies to every vicious indulgence, and such is any habit to be deemed which either impairs the constitution of him who is subject to it, or which implants the seeds of disease that, though latent in the parent, will be developed in the children. I know not, if tried by this rule, whether some things that are deemed innocent might not come under proscription. Even coffee and tea may be excessive stimulants to some constitutions, and their effect, in such cases, is to impair the natural tone of the nervous system.

Let this be laid down as a cardinal maxim, that *no refinement of mind resulting from a too susceptible organism, and no intellectual power obtained by the waste of bodily vigor, is on the whole desirable.*

XII

EXTERNAL RELIGION.

THE distinction between RELIGION and PIETY is sharply drawn by the Greek terms used in the New Testament.* There may be worship without pious emotion. The Greek sacrifice to the gods was without devotion—there was fear, but nothing of the filial reverence which is the essential element of the Christian faith. De Quinsey has well said, that all heathen worship assumed that the powers which had control of human affairs were lawless beings, who were to be propitiated because they were cruel. The evangelical spirit is, on the contrary, that of love joined with adoration.

There has, however, always been a proneness of the human mind to outward forms of worship rather than to the culture of devout affections. Even under the Christian dispensation, with all the light which has been shed upon the darkened minds of men, by the teachings and example of our Saviour, we find in every age the *ostentation of religion*

---

* Θρησκια and ευσεβεια—the latter of which is, in our version, translated *God-liness*. (1 Tim. iv., 8.)

more prevalent than the inward spirit of piety. Imposing ceremonies—the pomp and pretension of ecclesiastical dignitaries, and all the external show of religious worship, have arisen from the natural inability of the human heart to receive the truth in its simplicity, and to apply it as a discipline of life.

If there was a feeling of devotion to God—of penitence for transgression—if there was self-examination, whereby though our sinful nature is made more manifest, we are led to secret communion with God in confession and prayer, there would be no need of all the array of human intervention which the Church has established. The error of the world has always been of deeming God to be far off. The forms of worship in the Greek and Roman Catholic Churches have been akin to heathen practices. The invocation of saints has rested entirely upon the basis that they are nearer to God than we are, and that their intercession will avail when our own prayers would not be heard. The office of priest, so far as he is regarded as intermediate between man and God, is itself a superstition. It is true that the Jewish priesthood, as delineated in the Old Testament, may seem to have had something of the mediatorial character—but it is to be remembered that the proneness of the people to idolatry made an imposing form of worship, and a reverence for the ministers of religion, essential to the preservation of the national faith.

In the early age of Christianity, and especially after the subversion of the Roman Government and the dominancy of barbarian tribes, it may have been in like manner of vital consequence for the maintenance of the Church, that there should be forms calculated to **make an** impression upon the senses. Hence the pageantry of processions—the gorgeous ornaments of cathedrals—the stateliness of the superior clergy—the sonorous ritual of Church service—grand and even rhythmical in expression, sometimes modulated **to** the elevation and cadence of music, alternated with ejaculations and responses by the people.

Not that these things were devised by ecclesiastics merely in consideration of human weakness ; worldly pride was intermingled ; but there were many pious men in the Church who labored with sincere zeal for the advancement of religion. To them we **are** indebted for the devotional spirit that breathes **in** the ancient liturgy. **It may** rather be said that pride and ambition were permitted by Divine Providence to aid in the accomplishment of an ultimate design, which was to be permanent in its benefit, though the instrumentality was but temporary and gradually to be superseded.

Of the evangelical Churches of the present age, **or** at least of a considerable part of them, it may **be** said that they do not make much account of external forms, and it ought to be a subject of devout

acknowledgment that we have made so great an
advance upon what has preceded us. It is evidence
that, notwithstanding the long lapse of time and the
many superstitions that have prevailed, there **has**
been still a progress in the development of the
true Christian spirit.

There has never been a period in the history of
the world when sound rational **religion** has been
better established than it has been in our own time
and in this country. Perhaps **in** no former **period**
would religion have retained its hold of the **public**
mind if divested, to the extent it has been with us,
of all external pretension.

Such speculations should, indeed, be indulged in
with caution ; **we** are apt to make too free in our
judgment of the divine administration of the
affairs of this world.

When we take into view that religion is sustained
here, for the most part, by the voluntary aid of the
people—that without any other support than this .
it is almost universally acknowledged, however diverse the forms of worship or peculiar tenets, and
the diversity being in itself a proof of the vital
power which exists, since notwithstanding the differences, many of which are by no means trivial,
there is yet toleration and concert of action in the
great enterprises which properly belong to the
Christian Church—that our religion consists not
merely of a creed and of certain observances at

stated times, but it enters into our daily life, and to a great extent controls the conduct of men. When, I say, we think of these things, we may reasonably conclude that we are approaching nearer to conformity with the teachings of our Lord than those who have lived before. Imperfect as may be our performance of known duty, and much occasion as we have to confess our short-coming in conformity of life with the precepts of the Gospel, yet ought we also to be encouraged by the progress we have made and to render thanks to God for it.

The assembling of Christians for prayer—the many charitable associations in which different sects meet upon common ground in the spirit of Christian fellowship—the domestic training of children in a knowledge of religious truth and in conformity of life therewith—the easy, unaffected recognition of Christian obligations, even in the midst of everyday business—these are some of the characteristics of well-ordered churches in this country.

Their peculiar distinction is, however, rather for what they are not than for what they are. It is more the absence of hollow pretension and Pharisaic ostentation, which would impose upon the world by an appearance of sanctity that does not exist in the heart—claiming an exclusive, or, at least, extraordinary grace, and seeming to say to the rest of the world : " *Stand by thyself, come not near me, for I am holier than thou*" (Isaiah, lxv. 5) ; and on the

5

other hand, the absence of the fierce fanatical spirit, which claims to have spiritual discernment without the aid of reason—which has familiar communication with God, and despises all opinions of men that would restrain the riotous emotions of an insane **mind.**

The peculiar **trait of** professing Christians in this country **is, generally, a rational, consistent** character—not exaggerated beyond what **would comport** with human infirmity—not based upon *mere doctrine* set up as the standard of religion with **comparatively** little regard to conduct — but **intermingled** with, and giving tone to, the general habit **of life.**

But while we should have a **proper** sense of our advance in Christian culture, and should render de**vout** acknowledgment to God **for it, yet** it should not **be** forgotten **that** *the* ostentation *of religion* is *even now much easier* **than the** *realization of* it in **our** *hearts*—and that there **is** a proclivity to substitute outward observances, even though they be pain**ful** and tasking, in place of self-humiliation before God.

Many are thus unconsciously beguiled—finding it more natural and easy to submit to external penance, than to undergo self-examination, and to make secret confession to the Great Searcher **of** hearts ;—confession to a *priest* **has** been seen in former times to be a favorite commutation—so **the**

worship of saints and angels—anything rather than the unveiling of the heart before God. Is not **this** thinking of Him as unobservant of our inner life— *is it not the seeking of sympathy with human **in-firmity**, not for the purpose of obtaining divine **grace** sufficient to enable us to resist our natural inclination to evil, **but rather** a toleration of the propensity itself?*

We need to bear in mind the stern admonition of the Apostle, " Let **no one** beguile you of your reward in *a voluntary humility and worshipping of angels,* intruding into those things which **he hath** not seen" (Col. ii. **18), i.** e., the assumption of extraordinary sanctity. **Such** persons have always been in the world who will undertake to bear, not only their own sins, but also those of others—or, if no such profession is actually made, yet is there a tendency in the natural disposition of men to repose upon those who have obtained a repute for holiness.

Again, we are to guard against the self-deception of being satisfied with talking about religion, with **being in** association with Christians, and being engaged with them in charitable objects. Those are apt to *talk* most who give least evidence of their piety otherwise—I mean, talking by way of pretence, or to make a show of religion. **Of course, no** one can be said to talk too much who truly aims to impress **divine truth upon** others for their sake only, and not with a selfish or worldly motive. It was observed by an unlettered but practical preacher, that people of little religion are noisy—" he who has not the love of God and of man filling his heart, is like an empty waggon running down hill—it makes a great noise because there is nothing in it."

Association with pious people, attendance upon public worship, and a decorous observance of the usages which prevail where there is religious worship —these deceive many by a seeming likeness to the true Disciples of Christ, while there is, in truth, no inward affinity to Christ himself. Such men there are, alas, how numerous ! who live and die unregenerate, yet subject to an illusion that they belong to the people of God, or at least that they are as good as those who have made profession of their faith. And not alone the morally upright, but thousands who veil under an outward decorum the unsubdued passions of a selfish nature — envy, uncharitableness, avarice, sensuality. What shall be their lot in the great day of account, is fearfully revealed by our Saviour—what their failure in accomplishing even the purpose to which their life was devoted, is expressed in Ecclesiastes, " And so I saw the wicked buried who had come and gone from the place of the holy, and they were forgotten in the city where they had so done." (Eccl. viii. 10.)

XIII.

INEQUALITY IN THE CONDITION OF MEN.

HEREDITARY DISTINCTION—POVERTY—SERVITUDE.

INEQUALITY in society has always existed, and must continue to exist. It is a vain speculation by which is sought how to make the condition of all men alike. It was said by the Jewish lawgiver, not, perhaps, more in a prophetic view than upon profound observation of the character of our race, "The poor shall never cease out of the land," (Deut. xv. 11) ; and our Saviour said, "For ye have the poor always with you," (Matt. xxvi. 11.)

It has been a favorite theory with many, of late years, that society should be re-organized, and the hardships of a large part of the community are attributed to an arbitrary and unjust social constitution. If, however, we examine the origin of these inequalities, it will be found that they are inherent in the very nature of man. Oppression may be ameliorated, a greater degree of comfort may be enjoyed by the poor, but poverty must still exist, and privations continue to be the lot of a large part of the race.

While, however, we **may** deplore **the** general state, and feel a philanthropic interest in whatever promises relief **to** the suffering, yet we are **not, therefore, to** assume that all which seems **to be** hardship is unmixed evil. It would induce a doubt of **the** beneficence of the Divine Power by which **human affairs** are regulated, if so **large a** part of **the race have been** subject to a blight, without alle**viating** circumstances. **When we** find that distinctions have, in fact, always existed—that in **every** civilized nation there has been a class possessing wealth and power, and another class having few of the comforts of life, and coerced, by poverty and hopelessness of any better state, into a subservience to their superiors ; **add** to this, that in almost all **ancient** states, **and in** some even **at** the present day, there has **been,** and is still, **another** class held in absolute servitude — **having no** prospect of amelioration except by the slow process of emancipation, **and** then of citizenship after **a** long probation, **as libertini or** freedmen—we might be almost persuaded **that human** life had, **as** respects a large **proportion of the race, been** a failure. But taking a **large historical range of view,** we can perceive a great advance **in social order,** and **a** tendency toward a more equal **enjoyment of life.**

It **is** impossible, **with our** limited prescience, to measure the great **scheme by which** human capability is being **developed.** The lapse of centuries may be of slight account in the great succession of events. But while we cannot foresee what shall be the result of the progression which is now going on, and while there must necessarily be much that we cannot comprehend, for the reason that we cannot know what lies before us in the dim future, yet is there enough to satisfy any thoughtful mind, that the vicissitudes of human life are not fortuitous, or mere incidents of chance, but are subject to laws which at some period, perhaps far distant, shall become apparent in grand consistency.

In the meantime, it is an interesting speculation to inquire how far we are able now to judge of the probable purposes of those seeming irregularities which we have been too apt to judge as deflections from the order prescribed by divine wisdom ; or, in other words, to suppose, with the old heathen, that there is a perpetual antagonism between supernatural powers having control of human affairs, some of which are hostile and others friendly to the happiness of man.

The scope of speculation is, indeed, still within a narrow limit, yet conclusions may, perhaps, now be attained in respect to questions which have formerly baffled inquiring minds.

It is proper to add, that hypotheses which resolve the problems of human life by referring all that has transpired to a design had in view by Divine Providence, should by no means have the effect of super-

seding human responsibility, or impairing the enterprise of men in pursuit of their own plans.

It is a mysterious, yet unmistakable element, in
the regulation of the affairs of this world, that
while the will of man is free, and while it is often
perverted to the accomplishment of evil rather than
good, nevertheless, there is a superior force by
which the aggregate of human purposes and actions
are made to undergo a process like that of assimilation in the human organism, whereby a healthful
secretion is effected, though many ingredients, dele-
terious in themselves, or when not combined with
other things, may have been intermingled.

I propose to notice some of the principal distinctions in human society, and enucleate by some general considerations how far they have subserved any
beneficial end.

I. *Hereditary Distinctions.*—A class of men has,
in every nation, had superior rank by hereditary
right—in some instances involving the exclusive enjoyment of public offices, in others the possession
of patrimonial estates, the alienation of which is
restricted, in some degree, by law or usage ; or,
lastly, it may consist of honorary titles with an ap-
panage of personal privileges, to which may be
added wealth or official dignity, but there may be
a nobility without either of the last-named incidents.

Under a despotical government, the subjects are

reduced to a comparative level, as public office is bestowed at the arbitrary will of **the sovereign,** and a great estate, or noble lineage, is obnoxious to royal jealousy, which can admit of no rivalry with its own authority **or** splendor. Yet large estates, however subject to spoliation, must be to a considerable extent retained in families, through many generations, by natural succession. And even if the government should be maintained by military force, yet there must be chiefs whose rank **must, in** some degree, be transmissible. Martial dignity is susceptible of descent **as well** as civil honors.

In the ancient Greek republics, the whole body of **free citizens were substantially** equal in political rights, with only the exception of what advantage might belong to wealth. This was, indeed, consid**erable, as** in the isolation of those states, and the exclusion of all those changes which commercial enterprise and free intercourse with other **nations** might induce, family estates were perpetuated ; add to which, the great destruction **of life by war** prevented such **increase** of population as would diminish those estates by partition among many heirs. At Athens the division of citizens was made exclusively with reference to amount of prop**erty,** the right to certain offices being limited **to** the highest class ; but as all the citizens had **the** right of admission into it upon obtaining **the requi**site qualifications, and as the accumulation of

fortunes was more **frequent there than** in any
other state, **by the** effect of foreign trade, **it** might,
on the whole, **be** deemed **a** condition **of** equality.
Other than **the** distinction based upon hereditary
property, there was, **in** none of the Greek states, a
nobility, or if it existed in name, it was without any
power peculiar to the class, except **what** was de-
rived from patrimony. The Eupatrids of Athens
and **the** Heracleids **of Sparta** had **no** advantage
over **other** citizens, **save** in the respect which **might**
be voluntarily conceded to ancient families.

On the other hand, there was a vastly more **nu-**
merous body of men who were held in servitude,
either of **a** wholly menial character, as the Helots
of **Laconia, and** the household and predial
slaves **of Athens, or in** a subject state, without
political **rights,** as **the** Messenians **after** they had
been conquered **by** Sparta, and **so,** to some extent,
the colonists which **were** sent out by many of **the**
Greek states to occupy countries which had been
subjugated, which colonists were usually taken **from**
the poorer classes. The effect of slavery was to **re-**
lieve free citizens from mechanical labor ; they had
therefore **nothing to** occupy them except military
service, **the** administration of the government, or
the **pursuits** of literature **and art.**

The Athenians, who, **by** their liberal views of the
benefit **of** trade, were brought into more free intercourse with other countries than any of the Greek

states, were the most renowned for intellectual development.

Perhaps this is in fact in part attributable to the leisure resulting from the wealth which was acquired under their polity, but also, and in a greater degree, no doubt, to the natural genius of the people. The splendor of their achievements in the fine arts, eloquence, literature, administration of laws, and even in martial prowess, has been the wonder of the world. Brief as was the period within which this great development took place, we look back to it still with unsated admiration. If we inquire what benefit there was from the immense accumulation of slaves, especially at Athens, the answer can only be, that it left the whole body of citizens at liberty for public service. This was of the very highest consequence for the preservation of nationality, at a time when war was the ordinary condition of almost every state. An agricultural population could, therefore, not escape from servitude. The combination of trade with military enterprise was probably the highest phase of ancient civilization.

The Patricians of Rome were a nobility, and had, until the extinction of the Republic, a preponderance in the administration of the government. Participation in public offices was, indeed, obtained by the Plebeians after a severe struggle, but family distinction was always of great account with the Roman people. It was, indeed, possible for men of the lowest rank to attain the highest dignity, and even to become enrolled in the Patrician order, yet while there was enough to encourage the ambition of all in the arduous emulation for distinction, there was still a discrimination between the old and new nobility, and " novus homo" was an invidious term in aristocratic society.

The marvellous success of the **Romans, I** think, may chiefly be referred to the nobility, **with only** this qualification, that there was always a large body of Plebeians who were participants of the wealth acquired by conquest, and who constituted an important element in the military force. I mean that the common soldiery of Rome was not composed of a servile or dependent class, but had the self-respect induced **by** the enjoyment of political rights, and the prospect of honor as a reward for great services. Still the unrivalled firmness of mind which was in many times of peril exhibited by the Roman people, was doubtless in a more eminent degree the trait of the highest class.

The ambassador of Pyrrhus reported to **his** master, after an interview with the Roman Senate, **that** all the Senators appeared like kings.

After the appalling disaster at Cannæ, **it was** the Senate which sustained the sinking spirit of the nation. Instead of losing courage under the pressure of a calamity that would have overwhelmed any other nation, they cheered the drooping Commons by a vote of thanks to the surviving Consul for not despairing of the Republic.

It would not be within the proper limit of this brief sketch to enter at much length into an historical illustration of the proposition I am discussing. I select a single instance in later times, which will perhaps be more suggestive than a more minute analysis.

The personal freedom of the English nation is, in my opinion, incontestably due to its nobility, including, however, under this designation, the great land proprietors or gentry, although not noble by honorary title.

It was by the Barons that the great charter was extorted from King John, and the subsequent contests for the restraint of royal prerogative were maintained by them with the aid, in later times, of the class of landholders before referred to, who had become of some weight in the legislative body under the designation of the Commons. The burdens, indeed, of misgovernment fell heaviest upon the latter class, yet resistance would have been ineffectual but for the powerful intervention of the nobles. Even in the great revolution of 1648, when the Commons had become the preponderating power in Parliament, they received support from a large number of the nobility, and this circumstance was probably decisive of the struggle.

It can hardly be doubted that at no period could

the common people of England have escaped from crushing despotism, both civil and ecclesiastical, had it depended solely upon their unaided resistance. It may, indeed, with equal truth be said, that the great lords would have lacked a material element of power had they been without the support of a bold and manly commonalty. I think it is questionable whether what we now deem an arbitrary and inequitable rule of the English law, viz., the right of primogeniture, may not, after all, have contributed to the independence of character, which was a trait even of small landholders. It had, at least, the effect of keeping estates for a long period in the same family, which must have tended to maintain a sort of genealogical pride, and in a rude age, may have been the surest protection against the abject poverty which would naturally have followed a division of inconsiderable estates among many heirs.

II. *The Poor.*—It has been before remarked that there is no possible condition of society in which the poor will not be found. Instead of visionary schemes for the equalization of property, in contravention of the natural laws by which human enterprise is regulated, or the more fantastical project of relief by communism of property and labor, the true direction of philanthropy is to elicit a fellow-feeling of the better-endowed towards those who are less fortunate, and thus to keep up in the latter a self-respect, which is the only safeguard against debasement and its kindred vices. Great indeed is the change which in this respect has been wrought in Christianized countries. An enlarged application of the law which enjoined upon the Hebrews humanity to the poor of their own lineage, has, under the Christian dispensation, made compassion for suffering, wherever it may be found, a sacred duty. Even under the worst perversions of evangelical truth, by clerical ambition or fanatical zeal, it has never been forgotten. In the dark ages the monastery was an asylum for the destitute, and the giving of alms to the poor was the most common method by which men, who, under the religious views then prevalent, sought to make expiation for evil deeds. In our own time, the hospital, asylums, and various other charitable institutions for the reception of the sick and the helpless, and the reform of the erring, attest how deeply the principle of charity is fixed in the heart of the Christian community. Much still remains to be done. We have yet to learn more fully the extent of the sympathy which our Saviour exhibited to the outcast, and his compassion even for the depraved. In the light of divine truth, as illustrated by Him, an immortal soul, even in all the degradation of vice, is priceless. All the distinctions of human society sink into insignificance when compared with the intrinsic worth of such a being. Respect for what it is in its own nature, and tenderness toward the frailty by which the gifts so affluently bestowed are **perverted** from their proper uses, **should** supersede the Pharisaic ostentation which makes a **display of** charity **as** something entitled to praise, **while the objects of** the charity are **cheered** by **no words of** kindness. May we not also hope **that the time will come** when there shall be a higher **regard for the virtuous poor, when** rectitude of life **and unobtrusive** piety shall **be held as** constituting **worth rather** than the accidents of fortune. **I know, indeed, of** nothing which **more** surely indicates **the** tendency **of** any people to **the** destruction of all **the** essential elements of healthful, social life, than when there is a feeble discrimination of the qualities most conservative of the true interests of a community. **A** proper consideration for the upright, whatever **may** be their **condition of** life, **has** the effect of inspiring **self-respect in those who, in** adverse circumstances **and in humble life, are** irreproachable in their conduct. **The strength of any** nation depends **upon** the manliness and **virtue of** the lower class.

III. *Slavery.*—I know **of** nothing which **can be** alleged as an advantage of slavery, except the doubt**ful** benefit which may have resulted from it at **an early** period, when national independence could be maintained only **by military** force. **It** may then have been essential that the whole **body** of citizens should be ready **to bear** arms. **The** Romans, however, were renowned for their martial character before slavery prevailed to any considerable extent among them. I doubt if civil government gained anything by the exemption of free citizens from mechanical labor. The cultivation of literature and arts at Athens must be attributed to the natural genius of the people, as the same result of leisure was not exhibited at Sparta or Thebes. At Rome, not only common handicraft, but even the more liberal employments of architects and physicians, were exclusively appropriated to slaves and freedmen.

The slavery of Africans in this country must be held to be an evil with little or no counterbalance. It would be idle to discuss at any length the advantages alleged to be derived by the slaves themselves from intercourse with a Christian people. There will be some basis for this argument when the race shall be found to have a self-sustaining character, a proper test of which will be the concession that they are qualified for the enjoyment of equal rights as citizens of this country, or by the successful establishment of a government of their own in some other region. The former has been virtually decided against them—the latter is in process of experiment, with but doubtful omens of a successful issue. This last remark is to be understood, of course, as referring to a capacity for self-government in accordance with the laws of civilization. Even savage tribes have a sort of nationality and a rude administration of laws ; but the question is, if the African race,

which has acquired a knowledge of the institutions
and usages of an enlightened people, can, if left to
themselves, maintain in practical efficiency what
they have so learned, or whether they will retrocede
toward their original barbaric state. However this
may be, it admits of no difference of opinion, that
any speculative benefit of the kind supposed, is no
equivalent for the cruelty to which the slave is subject, and the vice which is the inevitable incident of
his condition.

Equally incontrovertible is the demoralizing effect
of slavery upon the master. The leisure which it
bestows has not in our Southern States been productive of increased intellectual development. On
the contrary, vigor of mind has preponderated in
that part of our country where mechanical labor is
performed by free citizens. Most of the scientific,
literary, and artistic productions which have given
to us a national reputation, have been educed in the
midst of the stir of commerce, and the industry of
artisans in the free States, not in the unfruitful
leisure of the South. It is but just to add that the
enervating climate of the latter should be taken into
account in this comparison. Can it, however, the
doubted, that under a more auspicious social system
a far greater intellectual activity might have been
the result ?

XIV

WISDOM OF THE ANCIENTS.

I THINK we very much overrate the actual knowledge which existed in classical antiquity. In illustration of this, I shall refer to the opinions entertained upon two of the most engrossing subjects of human thought—the one relating to the external world, and involving in the solution of questions to which inquiry thereof gave rise, a practical sense and patient observation—the other having to do with the nature of the soul, and calling for an entirely different function of mind, viz., reflection, or introversion of the process of investigation.

It will be quite beyond the scope of the brief illustration which I propose, to go into an elaborate detail. I shall chiefly cite from the two most learned of ancient writers, the elder Pliny, and Plutarch.

The theory of the World, or external Nature, is stated by Pliny substantially to this effect : The *Mundus*, or concave exterior which we behold, contains the sun, moon, and other planets, and is, in fact, the material limit of space. This vault of the heavens he supposed to be carried swiftly round, taking with it all the heavenly bodies, the earth being the centre about which **the** revolution **is** made, yet that these bodies had also another **motion,** which he thus explains : " The course of all the planets, and, among others, of the sun and moon, is in **the·** contrary direction to that of the heavens—that **is, to the left, while the heavens are** carried **to the right"** (this assumes **that we are looking** to the **north),** " **and** although **by** the stars revolving constantly with intense velocity, **they are raised up** and hurried on to the part **where they** set, **yet they are** all forced by a motion of their own in an **opposite** direction, and this is so ordered lest the air, being moved always in the same direction by the constant **whirling of the** heavens, should accumulate into one **mass, whereas now it is divided and** separated into **small** pieces **by the opposite motion of** the stars."

The Mundus **being thus placed, he** next proceeds to make some **curious** calculations **of its** dimensions, and of the distances of **the** heavenly bodies.   **The** opinion of Pythagoras is first quoted, who estimated the moon to be 126,000 furlongs* from the earth, and **from** the moon to the sun, double that distance ; **but Posidonius, it** appears, made **a more** liberal **allow-** ance.   **Yet** with all his fantastical conjectures, stated with **as much gravity as** if they had been reduced to actual measurement, **he nevertheless hit** upon one idea which is held by **scientific men** of our own times, viz., the limited **extent of the** earth's atmo-

---

* Somewhat less than 16,000 miles.

sphere. "There is a space," he said, "round the
earth of not less than 40 stadia" (the stadium being
125 Roman paces, or 625 feet), "whence mists,
winds, and clouds, proceed, beyond which the air is
pure and liquid, consisting of uninterrupted light."
From the clouded region to the moon he made
2,000,000 stadia, and thence to the sun 500,000,*
according to which the height of our atmosphere is
about five miles, and the distance to the moon
250,000 miles. Without settling whether there
might not be an error in this of a few stadia, more
or less, Pliny asserts, unconditionally and positively,
the principle by which the earth remains fixed in
the centre, amidst those counter motions of the
Mundus and the planets. Assuming that there are
four elements, which he says no one has doubted,
the highest being fire, the next highest being air,
these, with the other two (earth and water) are
balanced in this way, "the lighter being restrained
by the heavier, so that they cannot fly off, while
on the contrary, from the lighter tending upward,
the heavier are so suspended that they cannot fall
down ; thus by an equal tendency each of them remains in its appropriate place, bound together by
the never-ceasing revolution of the world, which
always turning on itself, the earth falls to the lowest part and is in the middle of the whole,   *   *

---

* The expressions used are " vicies centum millia," and " quinquies millia."

so that it alone remains immovable, while all **things** revolve around it."

It should be recollected that these, and similar hypotheses, are gathered by Pliny from a diligent reading of learned authors, among whom Aristotle had most favor with him.

We need not wonder at the crude results of the speculations of **so** many minds, when **we** see that even as late as the middle **of the 17th** century, the English Epic Poet represents Satan, after traversing chaos, as coming upon—

> " the firm opacious globe
> Of this round world,   *   *
>   "a globe far off
> It seemed—now seems a boundless continent,
> Dark, waste, and wild, under the frown of Night—
> Starless, exposed, and ever threatening storms
> Of chaos blustering round."

This is nothing **else than the Mundus of** Pliny, and is afterwards designated **by the Poet** as " a windy sea of land," and again as " as a crystalline sphere." So also he speaks of the earth as **a** " *pendent world*" *hanging on a golden chain.*

Yet Copernicus had, more than a century **before,** described the **true** planetary system, the motions **of** the heavenly **bodies,** and the revolutions of the earth. Such **is the** tenacity of ancient **error** that even Bacon could **smile** complacently **at** those " few carmen which drove **the** world about"—alluding to the great Prussian **astromoner and his** adherents.*

---

* See his tract " In the Praise of Knowledge."

The second of the subjects I proposed to illustrate, was ancient speculation upon the *Nature of the Soul.* The opinion of Pliny, which may be assumed to be the prevalent one in his time, is thus stated in his own words : " All men, after their last day, return to what they were before the first, and after death there is no more sensation left in the body, or the soul, than there was before birth. But this same vanity of ours extends even to the future, and simply fashions to itself an existence in the very moments which belong to death itself. At one time it has conferred upon us an immortality of soul, at another, transmigration, and at another, it has given to the shades below and paid divine honors to the departed spirits. As if, indeed, the mode of breathing with man was in any way different from that of other animals, and as if there were not many other animals, whose life is longer than that of man, and yet for whom no one ever presaged anything of a like immortality. ❋ ❋ ❋ How is *it* (the soul) to see, or hear, or touch ? And then of what use is it, or how can it avail, if it have not these faculties ?" He, therefore, dismisses all the hypotheses of a future state as mere delusions, and, indeed, as being undesirable that they should be true, inasmuch as it would cancel the chief good of human nature, death."❋

Plutarch, on the other hand, had a somewhat

❋ Nat. His., b. 7, c. 56.

vague idea of a perpetual **existence** of the soul. Human life being a visible state, and death being a return **to a** latent condition. " **The** birth, **or** generation of individuals, **gives** not any being **to** them which they had not before, but brings that **in**dividual into view—as **also** the corruption or death **of** any creature is not the annihilation or reduction **into mere nothing, but rather** the sending the dis**solved** being into **an** invisible **state."** So, **he says, the Sun, or** Apollo, **is** called **by the** names, **Delius** and Pythius, i. e., *conspicuous*, but the **ruler** of the infernal regions is called *Hades*, that is, invisible, and man himself was first called *Phos*, there being a perpetual desire **in** mankind of seeing and being **seen**—and **some** philosophers have held the soul itself **to be light, since** nothing **is** so insupportable to man as **obscurity.** *

Plato, **Aristotle, and others,** maintained the future existence of the soul, **but** when the doctrine **was** reduced to a popular form by explanations necessary to make it intelligible, it is apparent **that** a corporeal, or at least material state of **existence,** was what they had in view in all the various **modes** of exposition.

At most, **the soul was** but one of the higher of **the** four elements, that is to **say, either Fire** or Air, **or if,** as Aristotle **supposed, there was a** fifth nature or element, still it **was proximate and** like to the **others,** that is, material.

---

* Plutarch " On Living Concealed."

This is clearly expressed by Cicero : " Souls, when once they have departed from the body, whether they are animal (by which I mean capable of breathing) or of the nature of fire, must mount upwards * * * * or if it is that fifth nature" (referring to the theory of Aristotle) " still is it too pure and perfect not to go a great distance from the earth."

* * * No swiftness can be compared with the swiftness of the soul, which, should it remain uncorrupt, must necessarily be carried with such velocity as to penetrate through all this atmosphere, where clouds, and rain, and wind, are formed ; * * but when the soul has once got above this region and attained a lightness and heat resembling its own, it moves no more."*

If we look now at the representation of the soul after death, which, perhaps after all, best expresses the popular idea, we find it to be an *umbra* or shadowy image of the bodily form which it had in this life—retaining the same propensities and desires, and even something of corporeal appetite and sensibility. Thus Homer depicts the shades of the dead as gathering about Ulysses, at first voiceless and unconscious, but on drinking the blood of the animals which had been sacrificed, as acquiring memory and the power of speech.† It is worthy of remark that the poet locates them in a subter-

---

* Tusc. Quæs., lib. 1, c. 17.   † Odyss., lib. 11.

ranean region instead of the higher part of the
heavens, to which, according to the philosophical
theory above mentioned, they should ascend—the
cavernous entrance to this abode was fabled to be in
Cimmeria, a land unvisited by the sun, but Ulysses
did not go down into Hades. He merely made a
trench about the ground near the entrance, and when
the blood of the sacrifice flowed into the channel
thus cut, the spirits were attracted as if by a craving to drink it. It may be further observed, that
not merely were there these simulacra of the bodily
forms of men, but also of horses and chariots ; and,
again, the images of those slain in battle, or otherwise disfigured, were gashed or mutilated in like
manner.

A more lively picture is sketched by Virgil, in
which is reproduced, however, the same shadowy
imagery of the dead, with the difference only, that
they are seen in the interior of hell.

If, from these more prominent subjects, we should
proceed to an investigation of the knowledge existing in respect to practical science, a corresponding
meagre result will appear. The learned Athenian
or Roman dealt little with the merely useful mechanical arts. Medicine, physical science, so far as
it depended upon experiments, indeed, whatever
required patient observation, or manual labor, were,
for the most part, left to slaves and freedmen.
Poetry, oratory, history, and metaphysical philos-

ophy, were alone thought worthy of pursuit by the
free born (ingenui). If to this there should seem
to be exceptions, as Aristotle and **Pliny,** still while
conceding to them extraordinary zeal in the investigation of natural phenomena, I think they are
chargeable, and more especially Pliny, with a
hasty and indiscriminate admission of reports not
well authenticated, and a proclivity to take what
could be gathered from books, rather than to rely
upon personal observation, even when there was
opportunity for such a test.

It can scarcely be estimated how much the
literature of that age was shorn of fertility by this
neglect of observation. The scenery of nature seems
to be no element in ancient poetry, and if there are
occasional brief allusions to a remarkable locality,
to the variegating effect of morning or evening light,
the beauty of a landscape, or of some architectural
structure, we attribute much more to the poet's conception than probably was in his mind. The original
jejune picture receives coloring and proportion from
our own more cultivated imaginations. The Romans
especially were deficient in taste for natural scenery.
As often as the Alps were traversed in the itinera-
tions between Rome and Gaul, no description has
come down to us of anything more than the hardships
of the journey. It is related of Julius Cæsar, that he
occupied himself in crossing these mountains, on
his way into Gaul, with the composition of a gram-

matical work, "De Analogia."* So the ancient historians give us, in a rhetorical style, the narratives
of battles, and the speeches of generals, and of
orators (for the most part, however, composed by
the historian himself) but as to those details which
are essential for the understanding of the actual
condition of a people, the growth of literature
and arts, the increase or decline of population,
wealth and the like, we look in vain for them in any
but very late writers, as Suetonius and Plutarch,
and even these not dealing methodically with such
subjects, but only as mere incidents. As to the
Greek philosophy, it is sufficient to say, that it was
barren of any useful result, but merely exhibited
the acumen of great but misdirected minds. Macaulay pithily says, that it was *marking time* (in
military phrase) without making any advance whatever.†

---

* Humboldt has noticed this insensibility of the Romans. He allows to the
Greeks a somewhat greater susceptibility to impressions from exterior nature, but
still shows by large quotations that in them also the development was deficient.—
Kosmos, **v.** 2.

† Macaulay's **Review of** Montagu's edition of the works of Bacon.

XV.

THEOLOGY.

CHRISTIAN Theology has always been, to some extent, abstract or remote from common modes of thinking. This has arisen, in part, from the difficulty of the questions involved, but also in part from a defective discipline of mind in the greater number of those who have undertaken to act as instructors. At an early period crude speculations were inevitable. Evangelical truth was new to the world, and a rational mode of exposition could be acquired only by a long experience, in which human error would necessarily be intermingled, and which must be eradicated by oft-repeated elaboration. Thus, in the third century we find the mystical or symbolical interpretation of Scripture by Origen, and, although not in terms admitted as orthodox, by the later Fathers, yet it often re-appears in their expositions. Even Augustine, who had, in fact, great logical acumen in polemical discussion of

controverted doctrines, was, in his pastoral expositions of Scripture, fantastical at times even to absurdity.*

The scholastic divinity which was introduced by
Abelard, in the twelfth century, and came into
general use in the Western Churches of Europe, superseded the puerile method of commentary which
prevailed at that time, but was carried to the contrary
extreme. Instead of extracting a forced meaning
from texts of Scripture, there was but little reference to Scripture, but a subtle mode of disquisi-
tion upon doctrines was substituted, in conformity
to the artificial reasoning of the Greek philosophy.
Nothing could be more unintelligible to the uneducated people than this metaphysical theology, yet

---

* A single instance will suffice for illustration: In the homily upon John ii.
1-11, in which is related the changing of water into wine, he says, that the *six water
pots* represented the six ages of the world, viz., from Adam to Noah, Noah to
Abraham, Abraham to David, David to the Captivity, from the Captivity to John
the Baptist, from John to the end of the World. He then proceeds: "They contained *two or three metretæ* apiece. * * If he had said merely *three* apiece
our minds would have turned inevitably to the mystery of the Trinity, but neither
ought we, because he hath said *two or three* apiece, at once to draw aside the sense
from that application, for the Father and the Son being named, the Holy Spirit
is by consequence understood. Now, when the Father and the Son are named,
it is as though two *metretæ* are named, but when in them the Holy Spirit
also is understood three *metretæ*."

Justyn Martyr, in answer to the common argument of the Pagans from the
ignominious death on the cross, contended that the cross was not dishonorable—
that there are various things in nature which have that figure, as the sails and
masts of vessels, ploughs, spades, &c. and that what distinguishes man from
beasts is his having the sign of the cross made by the nose, and cited Jer. Lam.
iv. 20, "The spirit of our face, the Christ," &c. (in our version "the breath of
of our nostrils, the anointed of the Lord," referring, undoubtedly, to the King
Judah.)

it served to exercise the ingenuity of many acute
but misdirected minds.   It became popular in all
the schools of learning, and, during three centuries,
constituted the whole of religious instruction, except what came from an humbler source than professors of divinity and celebrated preachers.   The
mass of the people were, however, but little tasked
by the learned labors of the schoolmen, as they
were contained in a language understood only by
the educated class, which was comparatively a small
number.   Even the forms of worship were in Latin,
which was not understood by a large proportion of
the priests themselves, who officiated in the churches.
It is easily to be understood how little popular
religious knowledge could be derived from published
dissertations, or from preaching.   The former were
locked up in a language not understood by the
commonalty ; so also preaching, whenever there was
any display of learning ; and whatever there was in
the vernacular, was, for the most part, limited to
traditions of the Church, to laudation of saints, and
exhortations to penance, confession, and the like.
In fact, during the period above mentioned, preaching can hardly be reckoned among the instrumen-
talities of religious knowledge.   It was, indeed, the
aim of the clergy to maintain their own personal
dignity by inspiring a superstitious awe in the
minds of the people.   It would have been subver-
sive of the whole papal fabric to impart a *true*
knowledge of Scriptural doctrine.

The Reformation in the sixteenth century swept away a vast amount of superstition, and opened sources of knowledge till then never enjoyed in any Christian community. The translation of the Scriptures into the spoken languages of several countries, and the multiplication of copies by aid of printing, would, alone, have made an epoch in the progress of popular religious knowledge. But in addition to this, many of the most eminent Reformers sought, by clear and practical commentaries, to make Scriptural doctrine intelligible to all classes of men. The general character of the preaching, too, at that period, had an unction of fervid piety which moved the consciences of men, and induced profound questioning within themselves. This was the basis of the self-sustaining religious character which then was first developed, and has since become established in the more favored parts of Protestant Europe — a phase of character not limited to the learned or high-born, but, perhaps, even in a greater degree exhibited by the humbler class of society, and which is prophetic of the perpetuity of the change thus inaugurated, and of its advance to a more complete consummation.

While, however, evangelical truth was thus rousing the popular mind from its degradation to a consciousness of the immeasurable benefits which naturally flow from the Christian dispensation, when rightly understood, a system of Theology contemporaneously grew out of the discussions of learned men, which discussions were at first a necessity involved in the resistance and exposure of fearful heresies, and then continued for the settlement of diverse opinions among the Reformers **themselves. These were** naturally followed **by** Confessions of Faith in the different churches, conforming with the views of one or other of the great Theologians whom they respectively adhered to.

The utility of these discussions, **or of the formal** declaration of doctrinal belief, whether by **creeds or** catechisms, I do not **mean to** question. It doubtless was necessary, **for the** purpose **of making an end** of controversy, that the questions in debate should be in some way resolved, and that those who could not agree should separate and form distinct organizations. Two remarks will, however, I think, be found applicable to the doctrinal phase of Christianity in this modern development.

1st. In the violence of spirit **with which the** controversies upon doctrines were often pursued by the differing sects, there was an **absence of** that primary element of the Christian faith, in comparison with which all doctrines were secondary, viz., the *Charity* which "beareth all things." How strongly is the imperfection of our nature illustrated when **we see** Luther maintaining the real presence of **Christ in** the Eucharist, with a fierce intolerance **of the** opinion **of** Zuingle, that the bread and wine were mere

symbols. And so in succeeding times, the controversies of Christian sects have given occasion to
their common enemies, the Papist and **Infidel, to**
denounce their faith as having no' bond of attachment, but rather as productive only of rancorous
contention.

2d. Though there is occasion **for** rigid exactness
of phraseology, and even of metaphysical precision,
in expressing doctrines which have been the subject
of controversy, yet it is not the language proper**ly**
adapted to convey the information required for general religious instruction ; and especially is it inappropriate to the juvenile mind, which needs culture
rather of devout emotion than of intellectual sub-
tlety in detecting error. **The** formulary, indeed,
will still have its use for reference **when** any question shall be raised ; but **the** knowledge essential
for the daily conduct of life **is** far other than this.
It is a perverse inclination rather than any want of
scholastic acumen that is usually chargeable with
any errors of doctrine having much practical bear-
ing upon the Christian character. I have long been
of opinion that the *Westminster Catechism*, and
other similar compilations, are not well adapted to
youthful or uneducated minds, for the purpose of
imparting elementary religious knowledge. The
mode peculiar to these manuals is, indeed, wholly
diverse from that which is found **in** the Scriptures
themselves, and has the aspect of a complicated sys-

tem of doctrine, in which the abstruse and unintelligible hold the most prominent place, because these being most subject to learned controversy, must be most artificially explained, rather than as a clear and comprehensible exposition of what every man ought to know, and is capable of understanding.

With these general remarks I pass to the consideration of Dogmatic Theology, as taught by Professors and learned writers within the last century. It doubtless is essential that Ministers, who are to be the spiritual advisers of many men that have little time or facility for extended investigation, should be prepared to resolve all difficulties which may occur in respect to doctrine, and this will require greater erudition and more profound thought than might be supposed. The uneducated man, if he have an inquiring mind, will be as likely as one more educated to meet with doubts ; in fact skeptical opinions belong rather to the rude and worldly mind than to one accustomed to recondite reading. The latter, however, has his peculiar doubts, but they will generally relate to matters more remote— perhaps some speculative difficulty as to doctrine. All these a Minister should be competent to meet, and to give a satisfactory solution, or at all events, defend the right against any one who should persistently oppose it. This furnishes sufficient occasion for all the theological instruction that is imparted in the schools, at least so far as it bears upon all serious subjects of controversy. It does not follow, however, that a Minister is therefore to undertake to make all his people Theologians. The difficulties I have referred to will arise in a comparatively few instances, and they will be best disposed of when they actually occur. It would avail but little to preach a consecutive course of systematic divinity for the sake of forestalling all doubts which might by possibility occur to any of his people. His arguments would be little heeded except when immediately applicable to some existing case.

It has been the custom of the Scottish and American clergy to deliver elaborate doctrinal discussions from the pulpit—more so formerly than now ; a change has been induced by the increased activity of life, and the necessity of bringing religion to bear upon the varying phases of society, the stringent pressure of private business, and the startling incidents of public affairs. The efficiency of that mode of preaching, as a practical instrumentality of religion, even when there was less agitation of mind with the pursuits of worldly ambition, and a more stern estimate of the value of orthodox doctrine, may admit of doubt. The aged Christian or some strong thinker might find satisfaction in abstruse theology ; the light-minded of every age, but more especially the young, would heed but little the uncongenial discussion.

Austere discipline in the household may insure

the observance of all the forms of a religious life, without developing a voluntary and vital piety. The catechism may be made familiar as the spelling-book, and yet not elicit the reverent and deep emotion which evangelical truth, conveyed with judgment and affection, is calculated to foster.

So, in the Church, strict orthodoxy may be maintained, and yet a vitalizing power be wanting. Was not the Scottish Church orthodox in 1843, but had not practical religion so far died out that the old organization was abandoned by a large body of the people as having lost all efficiency for the spiritual advantage of its members ?

Again, did not the austere orthodoxy of New-England resolve itself, to a great extent, into antithetical Unitarianism ?—a transition which I have the charity to believe was the result of a sincere though misdirected effort to find some other element in religion than intellectual theology.

A single other remark will include all I have to say upon this topic. While scholastic divinity was reproduced in the pulpit, the relation of the minister to his people involved something of the superstitious respect which had in former ages been conceded to priests by an ignorant multitude. The reason was, that although knowledge of the Scriptures had become more generally diffused than in any preceding time, yet the abstract doctrines which chiefly constituted the themes of discourse were too far re-

moved from popular apprehension, and required too
much of artificial reasoning and recondite learning
to admit of being introduced into familiar conversation ; or if colloquial discussion occurred, it was
upon unequal terms. The minister was to the people as an oracle. The effect was two-fold, viz., to
induce spiritual pride in the former, and to repress
free religious inquiry by the latter. I speak, of
course, only of the general result of the system, when
carried out fully according to its natural tendency.
Many exceptions to the rule there were, no doubt.

But in our own time, the office of the minister
has become essentially different from what it was.
The activities of life have so far increased as to call
for direction or restraint by some other force than
abstract doctrine. Practical questions are constantly arising which must be resolved by the application of principles which have heretofore had too
little development ; and these must be illustrated
by purity of life—by renunciation of the vain objects of worldly ambition—by an unaffected earnest-
ness rightly to direct the weak and erring, and by a
tender sympathy with all who suffer, especially the
poor and friendless. It may be exacting too much
of the imperfection of our nature to expect these
divine qualities unalloyed in any one, nor would it
be just to prescribe this exemplification of the Christian character to the pastor only—the same rule
applies to all ; but his is the greater responsibility

who has the larger power of accomplishing good to others. And how shall he who undertakes to train, by holy discipline, " the sacramental host," fulfil that office, if he partake not of the self-sacrificing spirit of the Master whom he serves ?

XVI.

WAR.

ITS IMMEDIATE CONSEQUENCES HOSTILE TO NA-
TIONAL PROSPERITY—ULTIMATE USES IN THE
DEVELOPMENT OF LOVE OF COUNTRY, RELIGIOUS
FAITH, AND OTHER KINDRED VIRTUES.

———

WAR is one of the scourges by which we are ter-
ribly admonished that the world is subject to a
power superior to our own. Human passion is, indeed, the immediate cause, but there is something
in the outbreak of great commotions that leads us
at once to the superstitious credence of evil spirits
moving the hearts of men, or to the belief of a mysterious Providence, by which desolation and disorder
are sometimes allowed for purposes we may not now
comprehend. War, Pestilence, and Famine, are
alike in this, that they come often without premonition, and seem to have no discriminating retribution, inasmuch as the whole of a community are involved in suffering—the weak as well as the strong—
the pure-minded and pious as well as the vicious
and profane.

Yet are we not therefore to suppose that there is

no moral purpose in such events. Nothing can be more certain than that there is a control over all that occurs in human affairs. Mystery there is, indeed, as to the plan by which that control is directed ; but there is no room for belief that anything happens by chance, when we *observe* the grand convergencies of centuries into results, which in their turn become new forces for the accomplishment of other and greater ends.

I think the true religious view of War is, that like other great desolating powers, it is an instrumentality subject to the great moral Providence by which the world is regulated. We are unable to analyze the causes, or to forecast all the effects ; and it would be presumptuous to suppose that it can be stayed or controlled by the efficacy of weak human prayer.

It is my belief, that for reasons now unknown to us, a probation, varying, indeed, in its forms, but continually renewed in ever-changing phases, is necessary for educing what there is of good in our nature, and repressing the evil. .

War does, indeed, appear to be an unmitigated evil, when we follow out all its oppressive results ; yet is there also, in the midst of the lawlessness and disorder—the violence and vice—which are its ordinary incidents, a perpetual admonition to seek repose from these evils in something more mighty than human power.

I have formerly had a dread of war as the greatest calamity that could happen to our peaceful and prosperous people, and have shrunk from the thought of the evils incident to the array of large bodies of armed men, even if it should be for defence ; much more when we should be confronted by a *hostile* soldiery.   But a more extended reflection has brought the belief that even this dreaded desolation is not sent upon a nation without some counterbalancing agency for good.   **The** bonds of national affinity—of private friendship—of a brotherhood of all who are exposed to a common danger, become stronger ; and noble traits of character are generated, which would otherwise have been, perhaps, forever latent. Such are some of the moral uses.

In a religious view, there is even a greater consequence.   **Although** at first **we look upon the** conflict of hostile forces as wholly originating from the passions of men, and the result as depending upon the material strength of **the** belligerents, yet we learn in time to observe an overruling Providence **manifestly** displayed.   All that is fearful in human strife becomes revealed as subordinate to the same power which **created** man, and endowed him with capacity for good **or evil.**

The greatest military people of antiquity was the most religious.   Auguries were always taken before a Roman army was sent into the field, and an oath **was** administered to **every** soldier.   However perverted may have been their notions of religion, it is apparent that the grave, reflective mind, which was the distinctive character of the Roman people, naturally refers the issue of human events to a Divine Power. It is also equally true that a devout or reverential feeling toward a Superior Being, who has authority over all human forces, is itself a source of greatness in either an individual or a nation.

The martial enthusiasm of the *Saracens* was derived from their religious faith, which, though fantastical and puerile as a system of doctrines, had yet this one redeeming element, viz., the belief that all events are directed by **God.**

The *Crusaders*, in their turn, were inspired by a zeal which wrought out a higher degree of heroism than had ever before been exhibited in military annals, and great as were the destruction of life and the oppressive burdens imposed upon the people by the vast armaments which went out from Europe, there was, on the whole, more than an equivalent in the general effect upon civilization. A great advance in manly virtues, the softening of the hostile feeling previously existing between European states, the introduction of a more honorable mode of warfare ; in short, a change from a semi-barbarous condition to the modern forms of civilization, which are far in advance of whatever had been seen in any of the ancient nationalities. Such were some of the results inaugurated by these holy wars.

It is a remarkable incident of modern warfare that although larger armies are brought into combat, the loss of life is on the whole less in proportion to the numbers engaged than in ancient battles. This arises from a greater humanity to the vanquished. The Athenian army, taken at Ægos-Potamos, were all deliberately put to death by Lysander. The Macedonian army, that was defeated by Paulus Æmilius, was annihilated. It is related that 25,000 were slaughtered, while of the Romans there fell less than one hundred, showing that there must have been a massacre after the battle was decided.

Julius Cæsar, in a single battle with the Germans, under Ariovistus, slew 80,000 of them, and, in a campaign against two other tribes, destroyed, according to his own account, 400,000.

In striking contrast with such barbarity is the usage which now prevails among civilized nations, of taking prisoners all who cease to resist, and either exchanging them during the war or delivering them up after its termination.

When Hannibal, after eight months' siege, was about to make the final assault upon Saguntum, all that he would concede, as terms of capitulation, was that the inhabitants might emigrate to some other place, leaving all their property ; this not being accepted, the whole population was destroyed by the Carthaginians, or perished by self-immolation.

Equal atrocity was perpetrated **by the Romans** on the taking of Capua in the second Punic **war.** Of the nobler class, a great part was massacred ; **the** residue of the people were doomed to slavery.

In the wars between the Greek republics, **utter extermination or** slavery **was the** usual fate of the **vanq**uished people.

On the other hand, modern civilization has introduced usages of humanity. A law of nations **is** recognized, the violation of which in respect to one is deemed **an** affront to *all* civilized **states, and sooner** or **later, retribution is enforced against the** offending **government.** The **civil rights of a people subju- gated by arms are usually not interfered** with fur- **ther** than that the allegiance of the conquered peo- ple is transferred to the victorious power. There may be oppression in the exaction of revenue, or in military conscription, but the laws regulating private property, are for the most part **unchanged.** War **assessments are sometimes made** upon opulent **cities, when taken possession of by an enemy ;** but **even this is now generally held to be** an exercise of military **power hardly to be justified,** though it may still continue **to be** practised. When a **town** is **taken by assault, the** old barbarity still **prevails. No** degree of discipline seems to be adequate **to pre- vent** rapine and violence.

It is a singular antithesis in **modern warfare that a flying enemy is** pursued and slaughtered, while, after the combat has ceased, the wounded belonging to the vanquished, including as well those who have been stricken down in equal fight as those who have fallen without resistance in the rout, are attended to with the same care as the wounded of the victorious army.

It is related that the English cavalry, after the victory at the Alma (in the Crimean war), pursued the flying Russians, who had thrown away their arms for greater facility of retreat, and killed or mutilated the wretched fugitives—in the language of one of the officers, giving an account of the battle, "making heads and arms fly in the air." Yet, when the carnage had ceased, the vast host of suffering victims, friends and foes, were gathered up and transported to hospitals, a part of the Russians being sent to their own hospital at Odessa, with the understanding that they were to be deemed noncombatants until regularly exchanged.

XVII

SEMINARIES OF LEARNING—EDUCATION OF YOUTH.

IT is an interesting inquiry how far the studies prescribed in the education of children and youth have affected the development of individual character. It might be supposed that the course pursued in some countries, at certain periods, was calculated to suppress the natural elasticity of mind, and prevent its growth. Yet it will generally be seen that early education, however faulty, has still produced much the same result. Two reasons may be assigned for this: 1st. The years devoted to study are those in which the mind, like the body, is expanding into the luxuriance of maturity, and by its own natural accrescence, enlarging its vigor and capacity.

The remark of Adam Smith, as to the supposed benefit derived from sending young men abroad to travel (which benefit he denied), applies in some degree to the advantages which are in like manner attributed to the discipline of the school. "A young man," he says, "who goes abroad at seventeen or eighteen, and returns home at twenty-one, returns three or four years older, and at that age it is very difficult not to improve a **good deal in that** period of time."[*]

2d. **The larger** amount of knowledge, and of greatest practical value, is what is derived **from mutual** intercourse of men with each other ; the young, **especially,** have an advantage from coming into familiar acquaintance with those who are older, and **even** the association **of youth together,** tends to awaken **observation and** elicit thought.

I doubt if what is acquired **from** books, **in** seminaries of learning, is comparable in utility with **the** discipline derived from the co-aptation and mutual incitements of minds acting upon each other. This **is, of** course, with **the** reservation that corrupting influences are avoided, and these are many and perilous.

The education **of** the Athenian youth is commonly said to **have** been limited to Gymnastics and Music. The latter **term has** been understood literally by many writers, but it, in fact, included Poetry, Oratory, and even History ; and **we know that** in the time of Plato, Geometry was **an** cle-

---

[*] Smith's " Wealth of Nations," b. 5, c 1.

Mr. Jefferson strongly condemned the practice of sending young men from this country to Europe for an education. He wrote from Paris, **in** 1785, "An American coming to Europe for an education, loses in his knowledge, **in** his morals, health, and happiness. * * Who are **the men** of **most learning,** of most eloquence, most beloved by their countrymen, **and most trusted and** promoted by them ? They are those who have been educated among them, and whose manners, morals, **and habits,** are homogeneous with those of the country."—Randall's " Life of Jefferson," v. 1, p. 434.

mentary part of education, as may **be inferred from**
the inscription on **his** door.*

Yet there was, **in** the use of this term, something
indicative of great attention given by the Greeks **to**
harmony of sound. **Poets,** Historians, and Orators,
read their compositions in public. How much the
pleasure of the **ear was** consulted, may be judged
from **the** fact that there **was a** sort of rhythm even
in prose, and the Greek **taste in** this respect **was**
adopted by the Romans. **Thus, in** Cicero's criticism upon the Greek historians, the construction **of**
sentences is much more insisted upon **than** the merit
of the narrative. *Herodotus* (to whom he gives **the**
first place) he praises for eloquence ("**tanta est
eloquentia ut** me quidem magnopere delectet"),
next to him *Thucydides* is lauded for his condensed
and vigorous expression ("qui ita creber est rerum
frequentia *ut verborum prope numerum sententia-
rum numero consequatur.*")†

The *Universities* of **Modern Europe took** their
present form in **the 12th century.** They were originally *Theological* schools. **The name** is probably
derived from the **Latin** *Universitas,* a legal community or corporation, **and not,** as has been sometimes
suggested, because all sciences were taught in them.
Civil law was taught **at** Bologna **and** Paris, and **to**
some extent in England, **but** constituted **a merely**

---

* Ουδεισ αγεωμέίρικος ειϛίω.
Let no one enter who knows not Geometry.
† " De Orat.," ii. 13.

collateral department at Cambridge, Oxford, and
Paris. **The Inns of Court** in London, established
in the 14th century, superseded lectures upon law
at the English Universities. Latin was the language of the Church, and therefore was exclusively
taught in the schools ; during several centuries after
the establishment of universities in the principal
countries of Europe, **no attention** was given in public teaching **to** the vernacular tongue of either **of**
those countries. The services of religion were all
in Latin ; but those services were limited to **the**
forms contained in the Missal. Preaching, so far
as there was any, was addressed to the educated
class, and **was** a mere patchwork from the scholas-
tic philosophy. What exception there may have
been **was of a** very humble character. The Mendi-
cant Orders **of Monks,** who became **the** popular
preachers in the 13th century, **used** the ordinary
language **of the** people, but their preaching was
mostly **confined** to the puerile traditions concerning
Saints, and other superstitions of the Roman church.

The Theology of the **schools** had very little that
was adapted **to** the instruction of the uneducated.
It was, in fact, made up of subtilties remote from
any bearing upon common life ; and, **as** we should
judge, was equally profitless for the **proper** development of the youthful mind **that** was brought under
its discipline in the seminaries **of** learning. Yet it
constituted nearly all that was taught from the time

of Abelard till the Reformation in the 16th century.

Judging from the vast numbers of students that attended at the English Universities and at Paris, it would be inferred that there was no lack of ardor in the pursuit of the jejune dialectics which constituted nearly all that was obtained from a University education.

The Reformation opened new phases of thought. The Greek language was studied—the Scriptures were translated into several of the vernacular dialects of Europe, and preaching was addressed to the common people in a language intelligible to them.

Still, in the seminaries of learning, the old philosophy continued to be taught. The great change that took place was in the popular mode of explaining the Scriptures, which originated not with learned professors, but with pious minds, which had become thoroughly imbued with the principles illustrated by our Saviour.

Passing from that epoch to about the middle of the 18th century, we find in all the celebrated universities of Europe a largely disproportionate attention to the ancient classics. This may, indeed, be said still to continue in some of them, especially the English.

But within the last century physical science has become developed to such an extent as to constitute a department of knowledge requiring an undivided pursuit in order to become **thoroughly master** of it. Chemistry and Geology have **been created** within that period ; Human Physiology **has, for the** first time, been properly elucidated, and even Astronomy, which had been earlier brought into mathematical precision by Copernicus and Newton, had not, till a recent period, been reduced **to popular** comprehension.

Since all these **changes have been** accomplished, it becomes **an interesting inquiry how** far the **system** of **education has undergone a corresponding** modification.

I shall limit myself to one observation. Popular education has been transferred from the ancient seminaries to institutions of a later origin having a **more direct** and practical relation to the business of **life. The universities,** it is **true, are** largely attended, **but the** greater **proportion of** those who receive what may be called **a liberal** education, **that is** something more than **merely the primary** rudiments of knowledge, have **not been** inmates of a university. Public schools in England and on the continent, **nent,** furnish the means of education more to the advantage of those who are not destined to a learned profession, **than they** would, perhaps, obtain **at a** university—at **all events,** more cheaply. **In respect** to **one of the professions, viz.,** Medicine, **it has become the** more common method **to pursue only those** branches **of** science which **are strictly connected with it.**

The American colleges differ in some important particulars from European universities. Although there is a course of studies prescribed as the condition of bestowing a literary degree, yet there is a fair intermixture of all branches of knowledge which may be useful in any pursuit, and more especially of natural science. The ancient classics have by no means the same amount of attention that is allotted to them in the English universities. It has, indeed, become a popular theory in this country, that education consists merely in imparting practical knowledge. In one sense, this is, doubtless, well founded, viz., that the object of education is to prepare the mind for the business of life, whatever that may be. It is obvious, however, that the mere acquaintance with facts, whether in physical, political, or moral science, does not furnish all that is needed. There must be a habit of just reasoning upon those facts, and more practical wisdom is educed from careful collation of what is observed, even within a narrow range, with reference to results in common life, than would in general be elaborated from research, however far extended, that should not be subjected to any such test.

XVIII

REASON AND FAITH.

PASCAL'S doctrine was, substantially, that reason was insufficient of itself to inform us of divine truth, because our nature is corrupted, and while it remains so, is unimpressible by any exhibition of the true character of God, or of our **want of conformity** to his law. The remedy for this is only by a change of heart. When this takes place, the system of evangelical truth becomes at once clear and consistent. As to the method of accomplishing this result, there is an obscurity of explanation which strikingly illustrates the tenacity with which Pascal adhered to some of the errors of the Church in which he was brought up. The following extract will suffice as proof of this remark :

" Try, then, to convince yourself, not by the augmentation of proofs, of the existence of God, but by the diminution of your own passions. You would have recourse to faith, but you know not the way ; you wish to be cured of infidelity, and you ask for the remedy. Learn it from those who have been bound like yourself, and who would wager now all their goods—these know the road that you wish to follow, and are cured of a disease that you wish to be cured of. Follow their course, then, from its beginning—*it consisted in doing all things* **as** *if they believed in them ; in using holy water ;* **in** *having masses* **said,** *&c. Naturally this will make you believe and stupify you* **at** *the same time."* *

Perhaps *stultify* would better express what was meant, according to what **St.** Paul says, "If any man seemeth **to** be wise, *let him become a fool."* (1 Cor. iii. 18.)

Making due allowance for **the supposed virtue in** holy water, **and the** like, there **is a** truth **of some** importance involved. Infidelity grows out of a *disinclination* to **believe.** Wicked propensity **is** generally a characteristic **of a** scoffer at religion. Association with a profligate society may have the effect **of** undermining the principles of **a** better nature. **A** general laxity of morals in the Church, **and** especially in those who are placed over it for instruction of the members, and who ˼*should* illustrate their profession by an example of **piety, has,** undoubtedly, **had an** unhappy **influence** upon many minds that, **under** more **auspicious** circumstances, would have **gladly** received **the** truth as revealed in the Scriptures.

On the **other hand, whoever** has a natural inclination to devout feeling, or in other **words,** who **is** conscious of a want of **a** higher and holier principle than anything to be found in ordinary human im-

---

* Pascal's Thoughts.

pulse, will not **fail to find, in** the **system** taught by our Saviour, relief **from** all **perplexity.** The Christian faith to such **a** soul **is a resolution of** all doubt and **a** satisfaction of **all that it seeks** for.

Hence the German theologian, *Schleiermacher*, has proposed a theory of the essential constituents **of** evangelical belief, which was intended to avoid **the** assaults of modern **criticism upon** what he deemed external and **not** intrinsic **in** our **faith.** According **to this, we** find **the** evidence of **a divine** power when we are conscious of a sinful **state,** and obtain relief from it by coming into communion with the Church. The influence by which this **result** is obtained, is, however, not attributable to the Church itself, that is to say, the members thereof, **because in each** of them there is imperfection, but we must go back to the **first** source from whence comes **all** this power, that is, to Christ himself. This establishes fellowship between him and every true believer. Beyond this internal consciousness, nothing more, is absolutely essential— the miracles, resurrection, and all that is anticipated of the second coming of Christ, are not to be deemed integral **parts of** the doctrine upon which salvation depends.

However we may account for it, I think it may be assumed that religious **belief is, in** general, founded but in a small degree upon impartial judgment. Early education and social influence deter-

mine the faith of the greater part of mankind.
Even in the exceptional cases, as when under the
Roman Catholic regime, its doctrinal errors have
been rejected and its immoral practices repudiated
by individuals, it will be found that some counter-
influence was brought to bear which was more
powerful than ordinary association; a few instances
may still remain, but they are rare, in which a candid spirit of inquiry has attained the truth against
all opposing forces.

Thus it may be stated, as the general rule, that
custom, or, in other words, education and early
habit, pre-determine a man's religious creed.   Reason, which in this connection may be defined to be
an intellectual judgment, has comparatively little
to do with it.

It is true, that this early education may not insure a genuine faith in the religion taught, nor conformity of life with its precepts.   It is my conviction that even the lowest forms of Christian doctrine have maintained at least a pretension of in-
culcating a moral life.   The Jesuit, *Escobar*, might
furnish excuses for the ease of the conscience when
there was great proclivity to sin, yet evangelical
purity of conduct was, even in Jesuitical casuistry,
admitted as honorable, and there were not a few instances exemplifying it even in that double-faced sect.

The pure-minded *Pascal*, who was an uncom-
promising foe of Jesuitical hypocrisy, yet remained

in vassalage to the Church of Rome, and admitted
the supreme authority of the Pope over the consciences of men.   How is this to be accounted for,
except upon the supposition that education and
usage were more powerful than all the intellectual
vigor even of such a mind as Pascal's ?

The argument of Butler has now become almost
a popular aphorism, viz., that difficulties incident
to our pursuit of evangelical truth may be deemed
a probation—the proof being such as will not necessarily *force* conviction upon any mind, and yet
are sufficient for the candid and earnest inquirer
A curious subject of speculation is opened by this
hypothesis.   I think it necessarily results, that responsibility must correspond with natural capacity
and circumstances of life.   A fair inquisition for
truth, with a sincere desire to attain it, is all that
can be required.   But there must be something
more than a mere intellectual process, by which I
mean purely abstract reasoning.   A docile temper
of mind, a willingness to receive the truth, is of far
greater account.

It is my belief, that the difficulties which
worldly minds are most apt to meet with, or perhaps, I should say, objections which they profess to
be embarrassed by, in fact arise from the inherent
aversion of such minds to the truth.   There is no
real desire to be rightly informed ; doubts are
readily entertained, and then a long course of argu-

ment may be entered upon for the resolution of the objections, whereas they would, in most cases, vanish at once if there was a natural docility of heart. In general, it will be found, that skeptical views prevail only with such persons as are, by inclination of mind, opposed to the pure principles of evangelical truth. The teachings of our Saviour are at once received by the true-minded — by really devout and humble inquirers. They are rejected by those who desire to find justification for the pride of self-sufficiency, or for an immoral life. The skeptic is one who maintains the sufficiency of human reason without the aid of revelation, or who finds, in the enjoyments of this world, all that he craves, and is averse to being disturbed by any admonition of his disregard of the law of righteousness.

An unbeliever has, usually, a scoffing spirit, whence it may be inferred that there is depravity. A defiance of the opinions of other men, is not a trait of the wise and good. In the few instances where decorum has been observed by those who have rejected the Christian faith, there will still be apparent, upon close observation, some striking defects of character.

It may, I think, be deduced from these observations, that unbelief grows out of a natural aversion to religious truth ; and that, on the other hand, there is no serious difficulty where there is an inclination of the heart to God. The general principles of the Christian faith have nothing repulsive to a mind that sincerely desires to be rightly instructed.   As to more recondite **doctrines, there** may be discrepancies of opinion between **those** who are equally sincere ; but the larger charity **of** the Christian world, at the present day, allows such differences without intolerance.   I do not deem the separation into **sects as** schismatic.   **We** are wiser **than** Christians **were** at **an** earlier **period.**   So far as there is conformity of belief in respect to matters fundamental, **there** should be a feeling **of** brotherhood, and whatever **of** differences there may be **as** to other points, should be dealt with in a spirit of mutual tolerance and respect, and this, if I do not **much** misjudge, **is** becoming the prevalent feeling, **so as, perhaps, to** constitute an epoch in the history **of the Church.**

XIX

THE SUPERNATURAL.

POPULAR SUPERSTITIONS, OMENS, PREMONITIONS, &C.

BUSHNELL has brought out a theory which would seem to reduce what has heretofore been deemed to belong to the realm of mystery into a mere ordinary matter of fact; in other words, has brought the supernatural within the range of the natural. This theory is mainly founded upon a distinction between *things* and *powers*. The former are acted upon, the latter generate action. In the class of powers is placed the human soul, which has the capacity of originating action, whereby material elements are acted upon.*

It is assumed, as an important principle in this theory, that the *Will* is voluntary, that is to say, is under no constraint by the control of motive, and, on the contrary, may act against the strongest motive. This, however, appears to me a mere verbal refinement. It cannot admit of question that the will is determined by *something*—its action is not casual or merely the effect of chance, for that supposition would subvert the whole basis of character, and

---

* Nature and the Supernatural.

whether the action be brought about by a *good* reason or *weak* one—whether by some *extrinsic power* or by some *intrinsic proclivity*—call it what you will—there is still an influence which is the cause, and this is what is called motive. A mistake has evidently arisen from confounding motive with reason. A man may act upon what others would judge a *weak* reason, and yet, to his mind, it may be a strong one, whether attributable to an erratic judgment, or a perverted inclination.

But, passing this, we are led by the general proposition to the startling fact, that the soul of man is, itself, supernatural. It originates power, and it is connected with the spiritual and unseen, whence it derives an independent authority over the grosser elements, and requires individuality of thought and action.

The mode of communication with the spiritual world is not explained, but Mr. Bushnell rather rashly admits that the demonstrations of rappings, table-movings, and the like, may prove the presence of spirits, yet that from the nature of the communications made by them, they must be of a low order, perhaps belong to the genus of *evil spirits;* he, however, more sensibly remarks, that whoever is, by disposition, open to association, indiscriminately, with all sorts of spirits, will, undoubtedly, find companionship to his liking.

I have not, myself, so much faith in that phase

of spiritual communications. Not a single case has fallen under my observation, of a person professing to have such spiritual gift, who had any other intellectual superiority ; in fact, there has been, without exception, a defect in this respect, what, in common phraseology, would be called a want of balance. I am not prepared to believe that the highest function of the soul, viz., communication with superior intelligences, is most largely bestowed upon those who have not ordinary capacity to deal with beings of their own order.

But the subject scarcely deserves grave discussion. The irony of *Swift,* though irreverent, is perhaps, appropriate. I cannot help thinking, there was some basis for what he says of visionary devotion, or, more properly, the pretension to it— that *it is generally to be found in dilapidated earthly tabernacles, as houses are said to be haunted which are forsaken and have gone to decay.*

It is a different question, whether there may not be communications to a soul of noble endowments ; whether it may not make aerial excursions in the slumber of the body, or have nocturnal visions, or whether there may not be an invisible presence which, by a voiceless language, communicates to the soul suggestions which it finds impressed without knowing the source.

Popular superstition seems to have been tenacious of the belief of the intervention of *evil* spirits, as expressed by some of our poets. Thus, in Hamlet :

> "The spirit I have seen
> May be a devil, and the devil hath power
> To assume a pleasing shape ; yea, and perhaps
> **Out of** my weakness and my melancholy
> (As he is very potent with such spirits)
> Abuse me to damn me."

And in Childe Harold :

> "'Tis solitude must teach us how to **die,**
> It hath no flatterers—variety can give
> No hollow aid—alone man with **his God must strive.**"

> "Or it may be with *demons*, who impair
> The strength of better thoughts and seek their prey
> **In** melancholy bosoms."

As to *premonitions,* or supernatural revelations of future events, there are some curious traditional incidents that are well authenticated. Thus Lord Lyttleton, the younger (I mean the profligate son of the worthy and accomplished author of " The Dialogues of the Dead" and other valuable works), it is said was admonished by a spectre, supposed to have been the spirit of a former mistress, who had died broken-hearted in consequence of his abandonment of her, that he should die within three days. Hardened as he was in vice, the admonition was a shock which he could not rally from, and, having communicated the subject of his agitation, his friends gathered about him on the night that he dreaded, and, by way of precaution, having put forward the clock beyond the fated hour, he went to bed somewhat composed, but his serving man having been sent out on some slight errand, found, on his return, that his lordship was dead, holding his watch in his hand, by which it appeared that it was not yet twelve.

On the other hand, it is related of Cromwell, that a few days before his death he announced to his terror-stricken family that he had been informed, in answer to prayer, that he should recover. Such was the reliance, by all his attendants, upon the firm will and mysterious knowledge of the Protector, that the chaplain, in his prayer, said that he did not ask for the recovery of his master, as that had been already granted, but only that it might be speedy. Yet nine days after, the soul which had thought itself thus favored of God in its earthly aspirations, forsook its earthly tenement and all its grandeur, to enter upon that great future which it had vainly supposed had been opened to its finite understanding.

Shakespeare represents Romeo as having a presentiment of auspicious fortune on the morning of the day in which he was to hear the destruction of all his happiness :

> " My dreams presage some joyful news at hand,
> My bosom's lord sits lightly on his throne,
> And all this day an unaccustomed spirit
> Lifts me above the earth with cheerful thoughts."

It is my belief that such omens, of which there

is, doubtless, not an infrequent experience, are not supernatural suggestions, but rather illusions of the natural mind.

There is another sort of omen which is of greater account, but is observed more by others than by the person chiefly affected. Thus, in Scott's " Anne of Geierstein," a sudden change in the usual habit of the Duke of Burgundy is remarked by the English Earl of Oxford, as portending some great reverse in his fortune, and he refers to a superstition he had been taught in early years, that any sudden and causeless changes of a man's nature, as from license to sobriety, from temperance to indulgence, from avarice to extravagance, from prodigality to the love of money, and the like, indicates some great alteration of his circumstances for good or evil (and for evil most likely, as we live in an evil world) impending over him.

I have often had occasion to observe a singular self-deception in the minds of men who have accomplished any great success. They seem to feel an independence of all hostile forces in proportion as they have been fortunate in overcoming difficulties with which they have had to struggle. Yet how often do we see that such men, in the very acme of their self-complacency and security, are generally in proximity to some great reverse which they think not of.

Baxter has expressed this, in a religious aspect :

" It hath been long my observation, that when men
have attempted great works and have just finished
them, **or have** aimed **at** great things in the world
and have just obtained them, or have lived in much
trouble and have overcome **it, and** begin to look on
their condition with content, and rest in it, they are
then usually near to death or ruin."

XX

FEAR OF DEATH—PHENOMENA OF SUICIDE.

BACON has remarked, that he did not believe any one feared being dead, but only the stroke of death. On the other hand, Shakespeare represents the fear of death as involving far more than the mere fact of dying—

> " Who would fardels bear
> To grunt and sweat under a weary life,
> But that the dread of something *after death*—
> That undiscovered country, from whose bourn
> No traveller returns—puzzles the will."

There is a mystery in the fear which all men have of death while they are in health ; and yet how that fear is overcome as we draw near, in the course of nature, to what has been through life a terror. The weak and timid woman that could hardly endure to see the last throe of the dying, or to look upon the body from which life had fled, will at last, meet with composure the enemy she so much dreaded.

Montaigne has noticed, that the arguments of philosophers are of little consequence in preparing us for death. The peasant exhibits as much courage and endurance as one who has studied *Seneca's Morals* or the *Tusculan Questions*.

To suppose that the external circumstances of death—the groans, the convulsive breathing, the rolling up of the eyes, or the going out of life, are all that awaken our apprehension, would be assuming that a cheat had been practised upon us, as nurses are in the habit of terrifying children into good behaviour by stories of ghosts and the like.

Again, it is but a superficial view which would divest death of all terror because it is fearlessly encountered in the heat of passion, as in the pursuit of revenge, or under some strong impulse, as martial ardor in battle. The man who would expose himself to the peril of instant death for a vindictive purpose, will, in his solitary reflection, if he have escaped from the danger, be, perhaps, full of superstitious terror ; the soldier who might, in the clamor of contending foes, rush into the fight with impetuous daring, will, after defeat and flight from the conflict, be terrified by the apprehension of the enemy coming suddenly upon him.

It is a question which may involve a good deal of speculation, whether the *fear of death* or the *love of life* is the dominant feeling. We are so much under the influence of habit that we *naturally dread any sudden change.* An unexpected reverse of fortune will shake the strongest spirit ; overwhelming calamity, whether it be some of the strange revulsions that happen sometimes without a known **cause, or** by the aggression **of** pestilence, or the outbreak of war, will strike **terror** into **the** most self-reliant.

When we look at the change wrought by death, **we** inevitably feel a revulsion at the apparent destruction of all that we are accustomed to consider **as** constituting life.

The materialism of the *Egyptian* and the *Hebrew* has still some hold of the human mind, even in **our** advanced civilization.   The *mummies* of Egypt declare, better than any historic record, the popular belief that all which constituted life was indissolubly connected with the corporeal frame.   To this we may attribute the care with which **the** lifeless body was **preserved from** corruption, by inventive art, whereby, **at this day,** we have the opportunity of seeing those swathed forms which were the mansions of living souls some thousands of years since.

The prevalence of the same idea in the Jewish mind may be inferred from the pictorial language of **Isaiah, who,** though a prophet, **was** obliged to address **the people** in conformity with popular **opinions, and, therefore,** spoke of the dead as if the soul **still** belonged **to its** earthly **tenement (see Isaiah xiv.), and** that the souls of those who had lived and whose bodies had been deposited **in** their cemeteries, still had a sort of local existence in these habitations of the dead.

So the popular mind, even in our day, does not realize fully the existence of the **soul** separate from the body with which it has been, in all **its human existence,** connected. There **is a** general superstition, **though it may not be expressed, that** the inanimate body **has** still **some sort of** existence. There is a feeling of repugnance **to** the exposure of the unbreathing form to **the** cold air, with only a sheet for a covering ; **and so** when, with funereal ceremony, the body has been deposited in **the grave,** how natural **and** almost universal **is** the **feeling of** sympathy, **as if it had** still sensation **and might be** oppressed by the shutting **out** of air and **warmth.** Is there **not a** common desire to seek, **as a** resting-place for the **dead,** an exposure to the sun and the dry soil of some elevated ground ?*

I recently had occasion to observe the satisfaction which a poor woman, who had lost a husband and afterward a child, felt, upon the **opening** of the grave for the second burial, to find **that the coffin which** contained **the** body of **the husband was** entirely dry.

The feeling **of revulsion at the** annihilation of

* This popular feeling was expressed by Cicero in the *Tusculan Questions :* " Who is there that does not lament the loss of friends, principally from imagining them to be deprived of the conveniences of **life ?** * * * No one is afflicted merely on account of a loss sustained by himself. * * **That** bitter lamentation, those mournful tears, have their origin in our apprehensions that he **whom we** loved is deprived of all the **advantages of life** and is sensible **of his loss."**

life in the body, **has** been expressed by Shakespeare—

> " To lie in cold obstruction, and to rot ;
> This sensible, warm motion, to become
> A kneaded clod. "

Is there not an idea that the intellectual power
which has been expressed **in** the bodily organism, is
not wholly **suspended at death, as** represented in
the melancholy **verses of Lord Byron—**

> " The under-earth inhabitants, are they
> But mingled millions decomposed to clay,
>
>   *   *   *   *
>
> Or do they in their silent cities dwell,
> Each in his incommunicative cell ;
> Or have they their own language, and a sense
> Of breathless being, darkened and intense
> As midnight in her solitude—"

It is worthy of observation, that we dread any
sudden **change.** The changes that occur in life are
sometimes such as to paralyze the firmest minds.
How much greater is the transition from the present existence to one of which we know nothing.

The popular mind, even in this age of Christian
civilization, except so far as it has been enlightened
by divine grace, is, perhaps, little in advance of
what it **was** when there was no intellectual guide
but the **Greek** philosophy. **At** Athens and Rome
there was no settled **opinion as to the** existence of
the soul after death. A vague belief was, indeed,
entertained, that the soul passed into some other
form of life. *Plutarch* held **the** opinion that life

was a manifestation of the soul, and that death
brought it back to a *latent* condition.* Utter extinction of the soul was not a tenet of any sect of
philosophers, and yet was, perhaps, the actual belief of the common people. The repugnance to
such an extinction is well expressed by Milton—

> " To be no more—sad cure—for who would lose,
> Though full of pain, this intellectual being—
> These thoughts that wander o'er eternity."

It is my opinion, that the shrinking from death
is, in a great degree, attributable to the love of life,
which seems inherent in every human being. Strong
tenacity of life is the natural disposition, and it
yields only to an extraordinary disturbing force.
" Yea, all that a man hath will he give for his
life," is the expressive language of the Scriptures.

It was a remark of Montaigne, that even when
suffering excruciating torture from stone in the
bladder, which might have been thought sufficient
to induce a desire for death, yet that trivial circumstances still had a strong effect in keeping up a
wish to live : " the tears of a footman, the dispos-
ing of my clothes, the touch of a friendly hand, an
ordinary phrase of consolation, discourages and
melts me."

Yet in singular contrast with this tenacity of life,
is the facility with which, in certain emergencies,
suicide is committed. It is related, that when

* Plutarch " On Living Concealed."

8

Turin was occupied by the Spaniards, in the war between Charles V. and Francis I., scores of people destroyed their lives by casting themselves from the window, as a relief from intolerable oppression.

I have seen in some statistics of Paris, that suicide having been committed in the vicinity of the *Hotel des Invalides*, by voluntary hanging at the lamp-post, it was followed night after night by a repetition of the same mode of self-destruction, until the police authorities cut down the lamp-post and thereby put an end to this singular mania.

So it is said that in some villages in France, after a suicide by drowning, it has been usually followed by other like occurrences to an alarming extent, as if there was a kind of magnetic influence inducing it.

I think it will be found that a suicide usually occurs under a sudden impulse, rather than by deliberate premeditation. A girl drowns herself in the first paroxysm of grief for unrequited affection, or the cold-hearted villany of one to whom she had yielded her honor. The merchant who has been accustomed to the homage conceded to wealth, cannot endure the averted faces of his former sycophantic friends when his fortune is wrecked, and makes a violent end of his life under the first shock of his disaster.

Yet what an amount of misery is borne by many who may perhaps lie down at night with the *desire*, or if they have religious experience, with the *prayer*, that they may never rise again ; and still, week after week and month after month, is their inexpressible anguish endured without being tempted to seek relief by self-destruction.

This is a strong confirmation that suicide is ordinarily committed upon the first outbreak of some great misfortune, and seldom, if ever, after misery has grown familiar. We might have expected that Napoleon would not have survived his overthrow at Waterloo, but it would have seemed unnatural if he had destroyed himself at St. Helena, as a relief from the slow torture of a cancer of the stomach.

It is my belief, that greater courage is required to bear the ills of life in certain conditions, than to make an end of them by voluntary death. Thus Gloster is depicted by Shakespeare as seeking to throw himself from Dover cliff, under the agony he felt from the loss of both his eyes ; and exile from home—

> "O ye mighty gods,
> The world I do renounce, and in your sights
> Shake patiently my great affliction off.
> If I could bear it longer and not fall,
> To quarrel with your great opposeless will,
> My snuff and loathed part of nature should
> Burn itself out."

Yet after he had failed in the accomplishment of his purpose, reflection calmed his mind to endurance of his grief—

> "Henceforth I'll bear
> Affliction till it do cry out itself
> Enough, enough, and die."

The old Roman committed suicide with deliberation. It was with him a calculation what was required for the support of dignity before the world, and, therefore, self-destruction was deemed preferable to the enduring of any great reverse of fortune. Cicero speculates with some nicety of discrimination upon the comparative degradation which would constitute suicide a virtue, or rather which would make it a necessity ; for it appears not to have been held criminal to take one's own life, even without cause. He judged that Cato was under this necessity, while, perhaps, his associates at Utica were not —their lives had been less rigid in principle—but such had been his inflexibility of purpose, that it was fit he should die rather than behold the face of a tyrant.*

It is related, that when the captive king of Macedon besought Paulus Æmilius that he might be relieved from the ignominy of being compelled to walk in the triumphal procession, he was told by that general that it was in his own power to avoid it, meaning that he could destroy himself, and, after it appeared that he had not courage to follow the advice, all the sympathy of the noble Roman was turned into contempt.

Sometimes a misfortune, which was attended with no loss of honor, as some domestic bereave-

---

* De Off., 1, 31.

ment, was deemed a sufficient cause for suicide. The younger Pliny mentions the case of a friend, who, in consequence of an affliction of this kind, had resolved to starve himself to death, which, becoming known, he was with great difficulty persuaded to renounce his purpose.

XXI.

CHARACTER, AS DETERMINED BY CORPOREAL ORGANISM.

THERE are two methods of reasoning as to the origin of distinctive traits of character. The one is by assuming an exterior influence as a moulding process whereby a man becomes intellectually and morally what this environment is calculated to make him. In this theory there is, of course, taken into account something more than mere association in life, or accidental advantages, as of education ; it includes, also, the natural constitution of the body, aptitude of the senses, and the like. Thus, Mr. Locke, who denied all innate knowledge, even of moral principles, and held that it was developed by a process of reason, yet made the qualification that there were certain natural tendencies, which he defined to be inclinations of the appetite, not impressions of truth on the understanding ; in other words, a determination of a man's actions by certain physical wants. The other method is by supposing that there are impulses or inclinations in the mind itself, which, however counteracted by circumstances of life, will at some period appear, and which, though they may, to some extent, be restrained by outward and accidental influences, do nevertheless constitute the intrinsic character of the man. These impulses, it is said, lie out of the range of organic mechanism, constituting independent phenomena, though they may be aided or impeded by the structure of the body.*

The discussion of an abstract hypothesis would be of little practical value. What I have to say relates to some interesting facts, showing practically the influence of the bodily organism upon individual character.

I. It was an ancient belief that the faculties and propensities of the soul belonged to, or were located in, particular organs or parts of the body, which hence were symbolically referred to as representing these faculties and propensities. *Compassion* was located in the *bowels; Love* in the *heart; Understanding* in the *reins; Anger* in the *liver.†*

Plato located the understanding in the brain ; animosity in the heart ; sensuality in the liver.

This originated in the observation of the effect produced upon those organs by certain passions, as anger, which disturbs the liver, and causes a flow of bile ; *love*, or its opposite, *hate*, which respectively produced a distension or contraction of the

---

* Dr. Ideler has developed this theory with much originality of thought, in his treatise upon the " Treatment of Insanity."

† Thus it is said that Joseph's bowels did yearn upon his brother, (Gen. xliii. 30.) " Where is thy zeal and thy strength, the sounding of thy bowels and of thy mercies?" (Isaiah lxiii. 15.) My *reins* also instruct me in the night season (Psalm xvi. 7.)

heart ; *pity*, which caused a relaxation of the bowels ; or, perhaps, from observing that those persons in whom these organs predominated, exhibited the emotions in large degree which were thus located in them. The *reins* (kidneys), however it may be inferred, represented wisdom rather from their situation in the body, being retired and enclosed in fat, hence, symbolizing *secret* purposes, which are somewhat akin to wisdom.

There is some truth in these supposed relations. An intimate connection appears to exist between certain organs of the body and particular faculties or impulses of the mind, and they may affect each other reciprocally.

Analagous to this, is what has been recently observed respecting temperaments. *Sensibility*, or delicacy of mind, belongs to what is called the *nervous* temperament ; *good nature* to the *digestive*, (that is, when the nutritive or assimilating organs predominate, and this corresponds with the ancient hypothesis of the connection of pity with the bowels) ; *courage* and *enterprise* to the *arterial* or *sanguine* temperament ; *melancholy, jealousy*, and *revengeful disposition* to the bilious ; and when there is a quick sensibility, as if the nervous habit be, as it sometimes is, conjoined with the bilious, we see sudden outbreaks of anger, which are not transient, but long-lasting, which are known as peculiar to the bilious temperament, or which may, perhaps, be the cause of that temperament.

II. Another phase of the dependence of **the mind** upon the corporeal organism, **is seen in the** effects produced by the varying degree **of energy in the** bodily functions. **Activity of body** is derived from an impulse that **may be** considered **a** motive of the **mind, but there is also a** grosser kind of impulse, **which** may be discriminated as belonging to the corporeal system. Sensual desire produces a direct incitement of corporeal action, with which the mind has nothing **to do, except so** far **as it** may **be taken** as identical **with life or the vital principle. It is** mere **physical propensity,** instantaneous **and unreflective. When such a desire is strong, it resists the** control **of** the mind, and materially affects **the character of** the individual.

Health, or a sound state of **the body is** essential to the proper action of the mind. There **may be** intellectual power with a weak bodily frame, but it is probable that in such cases the brain and nervous **system,** which **are** connected more **intimately** than any other **part of** the bodily organism with the mind, are in a more healthful condition, **or have** more energy **than other parts.**

The mind **is** lethargic when the body is fatigued. Its rational **power is** suspended in sleep, and it has been supposed that the mind does not **in that state** think at all. Dreams **may be** limited **to a state of** imperfect sleep ; they are most **vivid in illness, or when the mind** has been much disturbed, and **seem** not to belong at all to a condition of perfect repose, or what is called a *deep sleep*. Those who sleep soundly do not remember any dreams, and, according to Mr. Locke's reasoning, it is only by remembrance that we can know that the mind has been employed in thinking.

III. *Connection of the mind with vital power or life.* Involuntary motion belongs to that mysterious principle of a living being which we denominate life. That it is the action of the soul, cannot be admitted by those who limit its action to what we are conscious of and can remember. But no one is conscious of the circulation of the blood, nor is the pulsation of the heart, though we may observe it and measure its rate, subject to our will. Again, it belongs to the brute creation, and we must admit the existence of a soul in the lowest forms of animal structure, and that it must be immortal like our own, if life or the vital principle is a function of the soul. Locke concluded that we could not prove the soul to be *immaterial*, but that this did not interfere with the assumption of its *immortality*.

The ancients believed the soul to be but a subtile or refined form of matter. " Vita continetur corpore et spiritu" (life is contained in the body and spirit), says Cicero, meaning by " corpore" the outward, corporeal structure. He doubted whether the soul was of the nature of air or fire.*

---

* Si anima est forte dissipabitur—si ignis extinguetur (if it is air, perhaps it will be dispersed—if fire, it may be extinguished) .—*Tusc. Quaes.*

Virgil describes it as like the wind, " par levibus ventis volucrique simillima somno."

IV. Recalling our attention more in detail to the influence of the corporeal organism upon character, there are some interesting facts, which I shall put together without much method.

Beauty of Person is an external power, but tends, perhaps, to intrinsic weakness rather than strength. When great natural gifts of mind are combined with it, they constitute the elements of a leader of the people—a man that will call forth popular idolatry. The great men who have been conspicuous in active life, and had great power over the minds of the common people, have generally been remarkable for dignity of person. When we read of the fondness of the Romans toward Pompey, who was, through all his life, a favorite, we hardly need to be informed by the historian of his princely grace. The majesty of Mohammed's person may account, in part, for the marvellous power which he had over a people that had never before been united by either force or popular art.

In a rude state of society, an imposing person is indispensable to the obtaining of popular favor. Would the splendid renown of Charlemagne, attested by the indissoluble blending with his name of the appellation of *Great*, have been attainable in the barbaric age which he illuminated, without his colossal figure and regal mien ? Something of this influence may be seen even in the higher civilization

of modern nations.   Do we properly estimate how
much the reverent admiration with which the American people looked up to Washington was due to
his imposing aspect ?

On the other hand, Bacon has remarked that beau-
tiful persons are seldom otherwise of much virtue ;
that they may be accomplished, but not of great
spirit, to which, of course, there are exceptions.   It
is, perhaps, more generally true that those who have
no such advantage, have a perpetual incitement to
effort, especially if they have otherwise much merit.
And if there be deformity, it is likely to induce
preternatural energy, either for good or evil. Hence,
it may result in a noble character, or, it may be, in
low-minded cunning.

The Irish orator, Curran, is an instance how personal defects can be compensated by conversational
power.   Lord Byron probably owed, in no small
degree, the intensity of his poetical effort to a slight
deformity ; something, too, may be attributed to
his want of social talent, and a feeling of inferiority
caused thereby, which roused in him an almost insane desire of distinction in some other mode.

Sir Geoffry Hudson, as described by Scott, is a
good illustration of the effect of diminutive stature,
without anything otherwise disagreeable. The Black
Dwarf exemplifies, on the other hand, the opposite
effect of disproportion, or anything repugnant to
our natural feelings.

Healthful functions of **the** *body* have more to do than is commonly supposed, not merely with energy of character, but with moral disposition of the mind. "To be weak is miserable doing or suffering." This, although represented to be spoken by Satan, is an axiom applicable to human life, though with some modifications. One of the essentials of greatness is a sound body—not necessarily great strength of sinews—but the physical force essential to the development of, and putting to practical use, the powers of the mind. There are instances which would seem to show that a strong will is superior to, or independent of, bodily infirmity, as we see exemplified in William III. of England, who was remarkable for indomitable resolution, and yet was, all his life, afflicted with ailments of body. Generally, however, energy of mind is weakened by bodily infirmity. There is less of enterprise and perseverance, perhaps, in part, for the reason that there is less desire. The applause of men is of little account to one whose hold of life is slight, and who is continually admonished of the precarious nature of all that he may hope for, even if he could obtain it. Wealth has comparatively little attraction for one who has no capacity to enjoy it, though it may derive some consequence by contrast with the discomfort of poverty.

Strong passions are the nerves of the mind, and these are not ordinarily developed when there is little opportunity of indulgence. Happiness is chiefly dependent upon social sympathy, and therefore a man who, by ill-health, is precluded from much intercourse with others, or whose thought is continually directed to his own suffering, is shut off from much of what constitutes the ordinary enjoyment of life. One modification, however, is sometimes seen, viz., that as the range of sympathy is curtailed it may become intensified within a circumscribed limit. Thus, there is sometimes exhibited great sweetness of disposition and pure affection by an invalid; but this is when disease affects the grosser part of the bodily organism, producing pain or lassitude, but leaving free the more spiritual part of the system, which is in nearest connection with the mind.

Another phase of bodily infirmity is its religious aspect; perhaps its tendency is to produce a higher order of piety. How many hours are there that one who is subject to habitual pain must suffer in silence? The sympathy of friends cannot be always tasked. Even the affection of woman, which is proverbial for its endurance of many trials, will fail, or be seriously impaired where an invalid is subject to ailments of long and unintermitting continuance, especially where there is a breaking down or gradual wearing out of the faculties of mind.

And were it otherwise, it would be an ungenerous abuse of friendship to make it bear our burden, when we have it not in our power to perform, in our turn, the same kind offices to others. At such times, how lonely seems the world—how vain all human solace—then does the soul look through the gloom to a higher source of comfort, and find, in a clearer faith and brighter vision of spiritual things, the mysterious communications which the stricken spirit often receives from the invisible world, a consolation which, to those who have not had the like experience, is wholly inexplicable.

There is, however, a counter color, which should be noticed. In a state of health not wholly disabling a man from active business, and yet interfering with consecutive or long-continued exertion, causing at least exhaustion as a consequence of toil which it is necessary to undergo, there will inevitably be a low state of feeling, which, with the increase of ailment, will become morbid. As religion has to do with feeling as well as conduct, it will, in such a case, be likely to take a sombre hue. Infirmity, whether it be the effect of age, or of disease, may induce a pious tone of mind, and a praiseworthy resignation ; yet the religious experience of those who have great trials of this kind, has a melancholy, or at least, uninviting aspect, to those who possess a robust habit of body. There is ofttimes a want of charity in prescribing a particular phase of religious character, as a standard for all. Serenity of mind, humility and sympathy with distress, are developed in those who have suffered great trials, and these are admirable qualities ; but they are not to be looked for in an equal degree in those who are engaged in very active duties. Each should respect in the other whatever of merit is peculiar to him, and should mutually aid each other. The Christian, serene in bodily weakness, may impart something of his peace of mind to the robust and vigorous, who are more exposed to the disturbing influences of the world. Nor should the latter undervalue the subdued tone of those who are by bodily weakness comparatively secluded from the conflicts of life, but communicate to them something of their own buoyant hopes and energy of resolution.

XXII.

SELF-RENOVATION;

OR, METHODS OF REPAIRING WASTED ENERGY OF MIND.

IT may be a question whether there is exhaustion of *mind*. Possibly it may be only the corporeal agencies which suffer waste by mental effort. Be this as it may, it is certain that severe study, or any long-continued tension of the mind, seems to call for some alternation of bodily exercise, or some indulgence that may be denominated sensuous in distinction from the intellectual.

Such indulgence may have many gradations from the gross to the refined, but the *principle* is in all cases the same. Sir Isaac Newton found relief from thought in the use of tobacco, which he smoked to excess, and this seems to be the general habit of German scholars. Other more innocent modes of recreation are mentioned, as the practice of *Bourdaloue*, to dance to his own music upon the fiddle, at intervals, while writing his sermons, and that of

Montesquieu, who relieved himself, when wearied
with the composition of the Spirit of Laws, by reading the Arabian Nights, although there is some evidence in his Persian Letters that he had a taste for
grosser pleasures. Indulgence in wine, opium, or
sensual love, may be set down as the gratification of
an appetite rather than as an aid to the mind,
though it is related of Blackstone that he habitually had a bottle of wine by him when writing his
Commentaries, and there is unmistakable evidence
in some of De Quincey's writings that they were
produced under the influence of opium.

The more refined pleasure derived from music, or
the sight of paintings, sculpture, scenes of nature,
and the like, may be considered as appertaining to
the mind. Taste in the fine arts is an intellectual
accomplishment, though in actual exercise, involving
perception by the eye and ear.

There is also a relief to the mind in the alternation of different subjects of thought, and it will be
found that most men involuntarily recur to some
habitual or favorite subject. It was a significant
fable of *Antæus,* that he recovered his strength
whenever he touched the earth, which was represented to be his mother, and was overcome only by
being lifted up from it. So there are certain thoughts
which we fall back upon to gather up our strength,
when wasted by exhausting labor, or the distrac-
tions of life. It is this, indeed, which maintains

individual character. Without it, the many would
be fused into one, a few paramount spirits giving
character to all others, so as to produce a common,
undistinguishable likeness, or at least to wear away
whatever is peculiar to any one not having a strong
will.   But there is a time for reflection—memory is
busy with the past—early thoughts and resolutions,
and tender associations, come back to us.   It is a
kind Providence that such reminiscences have a tendency to soften down the asperities of our nature,
and to renew whatever there is in us of good.   The
loved ones who are dead are still guardians of our
honor and happiness ; in our solitary hour they rise
before us with a power like what would be felt in
their actual presence.   And if among these visions
of the mind there is anything that reminds us of
wrong which we have done, the feeling induced is
not altogether of painful self-reproach, but rather of
chastened sorrow, leading to or confirming a purpose
to expiate the past by a better future.   Happy is
he who has learned where his strength lies, and
what are the recollections which most sustain his
good resolutions, for he is then armed against temp-
tation and evil influences, and though he may sometimes yield to them, he will, at the first interval
of thought, recover himself.

　　But if there are those who are able to resist this
potent agency ; if the heart is so hardened as to be
unimpressible by the memories of by-gone years,

we know that it is a spell which will some day be broken.*

An antithetical effect is thus seen—the external or direct influences in the action of life, alternating with the internal or reflective ; in other words, an oscillation between sympathetic impulses and the restraints of thought ; and the observation of this has led many into a revulsion against worldly associations, and a settled seclusion into the privacy of their own thoughts—a course akin to what Foster speaks of as his experience, " a feeling of revolt when he found himself coming into anything like intimate, confiding kindness with any except a very few."

This course of thought furnishes confirmation of the psychological theory, that character is made up of original impulses, and that the proper balance is maintained only by the counteraction, or at least restraint, of one impulse by another, thus making an antagonism of motive powers, which in their joint action, maintain a general harmony, if one does not obtain an undue preponderance over another. But whenever the action of one becomes excessive, or out of proportion to that of others, the proper

* Foster has described this state with gloomy energy : " Does this dead stillness of conscience appear an awful situation ? Why does it so ? Because we foresee that it will awake, and with an intensity of life and power proportioned to this long sleep, as if it had been growing gigantic during its slumber.   *   *   It will awake—probably in the last hour of life—but if not, it will nevertheless awake ; in the other world there is something that will awake it—*at the last day.*"

balance or equilibrium is subverted. Then comes what may be properly termed insanity—an unsoundness of mind. The remedy for this is by restoring the healthful functions of the powers which have for the time been crowded out of their proper office. Insanity is, in fact, the concentration of mind upon one subject, or perhaps we may more properly say that this is the incipient stage ; when the balance is once lost, there is nothing but chaotic confusion in all the elements of thought. It becomes obvious, therefore, that restoration to a natural state is to be accomplished by bringing back the mind to the multiform processes which, by a law of our nature, are essential to its healthful condition.

Following out this theory of various powers, each useful in its proper action, but dangerous if it shall obtain undue mastery, thus symbolizing a free, civil government, in which there are many co-operating powers, each constituting a check upon the other, we can see something of the nature of the probation to which human life is subject.

The hypothesis of Comté* has a basis of truth, viz., that energy of action is derived from instincts which, in every stage of civilization, preponderate over the intellectual faculties. The latter "being naturally the most energetic, their activity, if ever so little protracted beyond a certain degree, occasions in most men a fatigue which soon becomes un-

* Comté's "Positive Philosophy."

supportable." Intellectual activity is, therefore, to be incited by impulse from the *lower but stronger propensities*, and the nature of man is elevated in proportion as he is moved by the *better propensities*, but in social relation, the lowest, or most sensual, have, in fact, the ascendency.

The moral growth of an individual is in correspondence with the influence which the nobler principles of his nature have upon his life ; the advancement of society is, of course, measured by the individual character of its members. Thus we see that we are, by our natural constitution, impelled to activity by our grosser proclivities, and that we are restrained from excess by intellectual powers, the culture and preponderating exercise of which constitute all real advancement in civilization.

XXIII

GOVERNMENT AND LAWS, AS AFFECTING THE GREATNESS OF STATES.

THE hypothesis that national character is originally determined by its laws, is a fallacy. I think it clear that government and laws grow out of the customs of the people, or rather that they are coeval with early customs and habits of thinking. The common version that the laws of some of the celebrated ancient states were devised by some lawgiver, and that the subsequent character of the people was derived from these laws, is a mere illusion. The most that was done by any personage of this kind, was merely to bring into harmony the laws already existing, and interpose some checks against sudden revulsions.

It has been shown by Bishop Thirlwall that *Lycurgus* did little more than re-enact the laws and constitution of government already existing. The assembly of the people, the *Gerusia*, or Senate, the two kings, and almost all the usages prescribed by him, were already known ; some modifications may have been established, but the chief thing done was to reduce to a systematic form customs already existing, and to render them permanent by religious sanction.

So, too, at Athens, the assembly of the people, the Senate, the Areopagus, and the laws relating to civil rights, existed before the time of Solon. What he did was merely to settle certain questions which were in dispute between the common people and the aristocracy, and to establish permanently the administration of the laws, which had before been oscillating according to the ascendency of one or other political party.

The history of Roman laws has been much misunderstood. The common version is that the Decemvirs compiled a code of laws, for which purpose, it is said, they visited Athens, and hence it has been supposed, that they transplanted what they enacted, from other states. But it is certain that the Twelve Tables are not made up from Athenian laws ; in fact, it is intrinsically evident, as well as historically shown, that they were made up from the customs of the Romans already existing, and that they constituted but a small part of the unwritten law which was then in force, and which was recognized and acted upon even after the Twelve Tables had been adopted.

I think it will be found, that in every state, ancient or modern, there has been a customary, or unwritten law before there has been a compilation in the form of codes or statutes.*

The common law of England was, for centuries, merely traditionary, and even at the present time, has no authentic record, except in the decisions of courts, which are fragmentary, and have been reduced to system only by the compilations of commentators. Statutes have merely modified the existing law, or declared it when it was doubtful.†

As to the English Constitution, it was wholly the growth of public usages. It derived little or nothing from theory. There was never, in fact, any formal reduction of it into a written or statutory form. The Magna Charta, and various other statutes, merely restrained royal tyranny, or feudal abuses. In the latter part of the reign of James I., and thence on till the revolution which commenced in 1640, we find, for the first, an appeal to principles of natural right. Yet, even then, the laws were not methodized into the form of a code or written constitution. The same remark applies to the revolution of 1688. Even at the present time the structure of the British government, and the political rights that are secured by it, are understood, only or chiefly through traditionary evidence

---

* This course of reasoning has been pursued more at length in the Article "Code" in Appleton's "Cyclopædia," which, I may properly say, lest I should seem to have borrowed from it too freely, was written by me.

† See article "Common Law" in the "Cyclopædia."

of usage, reduced into some method by elementary writers.

I think there has been no instance of a complete form of government established by legislative enactment until the experiment was made in 1789, by the North American States. The precedent was afterward followed by France, but all the constitutions set up in that country during the revolutionary period were ephemeral—they successively and rapidly were obliterated, giving place to a regime founded upon the practical exigencies of the times, which called for military power rather than a theoretical partition of civil rights. Defence against foreign aggression, in the first instance, and the maintenance of national eclat subsequently, constituted a greater pressure than all the inequalities and abuses, relief from which was the first motive power that inaugurated the revolution. A memorable revision of the laws, as affecting private rights, was, indeed, accomplished during the reign of Napoleon. It is, however, a mistake which, perhaps, has become general, that the French code was a new or original system of laws ; it was no such thing. It was, on the contrary, a compilation by able lawyers, from all the different codes or systems of laws existing in the different provinces of France, and merely introduced a uniformity throughout all the departments, with some modifications suggested by the experience of the jurisconsults who were charged with the revision.

It is scarcely necessary to notice the crude and abortive attempts, in Mexico and South America, to establish institutions or systems of laws in imitation of the legislative action of the North American States. There had been no discipline of the popular mind by tradition and usage. The result has, consequently, been a failure. There has been the form of a constitution and the semblance of laws, nominally securing equality of rights, but there has been no sufficient popular intelligence, or firmness of character, to insure any substantial benefit from the constitution and laws thus enacted. Military rule has been paramount ; legislative enactments are powerless ; there is, in fact, no freedom, except in the formal declarations of laws, which are wholly disregarded.

My present purpose is, however, not so much to trace the origin of laws as to deduce their effect in whatever mode they may have been constituted upon the development of natural power.

There is some exaggeration as to the effect of the form of government upon individual rights. A despotism, it is true, may arbitrarily supersede, or abrogate, all private or individual immunities. In ordinary course, however, this is not done. The chief injury done by an arbitrary government is the leaving too large a discretion to unprincipled subordinates. The wrong, however, that may be done by the inferior official is not for the benefit of the

general government ; on the contrary, it may be set
down to his own private emolument, and this may
admit of a more summary remedy under a despotism than under a free form of government. The
difficulty under the former is mainly, that proper
information cannot be obtained ; at least this is the
case where there is corruption among the officials of
the central government. On the other hand, if there
is integrity in the executive of a despotical government, a wrong can be readily corrected. Under
Domitian, it was useless to appeal to the Emperor
from an unjust decision of a proconsul. The
Emperor Trajan was ready to hear, and decide
justly, all complaints against unfaithful magistrates.

In reading the correspondence of Pliny with the
Emperor Trajan, I was much impressed by the
quotation of rescripts of the Emperor Domitian as
authority. The explanation is, that although within the immediate range of the flagitious passions of
that monster, no one could be safe—in fact, the administration of laws was wholly perverted—yet, at
a distance, where the emperor could act only by
deputy, the general course of affairs was according
to precedent, subject, of course, to exceptional
abuses, for which there was no remedy. I have the
impression, however, that the wrongs perpetrated
by the deputies of the emperor were less than what
were committed under what was called a free gov-

ernment,* but which was, in fact, nothing but the rule of a lawless populace.

It is an anomalous circumstance that the most acute reasoning upon civil rights was educed under the rule of the most tyrannical despots that ever ruled the Roman Empire. *Papinian* lived in the reign of Caracalla, *Ulpian* under Elagabalus. The Institutes, Digest, and Code, which constitute what is known as the civil law, were compiled under the order of *Justinian*, who had as absolute authority as any Asiatic despot.

In the modern states of Europe there is a great difference in political rights ; in some of them, as in Russia, Austria, and other states, there are no political rights ; that is to say, the common people have no voice whatever in the election of rulers or the determination of their measures. In France, at the present time, there is hardly less of absolutism. But, nevertheless, in all countries, there is an administration of laws as affecting private rights according to certain fixed rules.

The evil under a despotical government is the difficulty of getting redress for any injustice done by an administrative officer ; yet if the ruler has integrity, there will be more speedy retribution where there is absolute power, than there would be under a free government. There can be no doubt

---

* I refer to the form of government shortly before the breaking out of the civil war, in which Cæsar gained absolute power.

that wrongs were committed by Roman magistrates with greater impunity, in the time of Cicero, when there was liability to public accusation before the comitia of the people at the instance of any one who chose to undertake the prosecution, than could have been done under Trajan or the Antonines.   Yet it is certain that the imperial government had the general effect of diminishing population and depressing enterprise in the provinces.   This may be attributed, 1st, to the noxious influence of wicked men, who wielded the imperial power, at least, as large a proportion of the time as men of better character ; 2d, to the impossibility that an individual could effectually supervise the vast and complicated affairs of a great people ; he would, by necessity, trust the greater proportion of what might come before him for supervision, to subordinates, and these, of course, would be subject to corrupt influences.

The chief difference between a despotical and a free government, so far as respects the enjoyment of private rights, I take to be this : in the former, redress of grievances depends upon the will of a single person, who, by necessity, must depute others to attend to the greater part of the complaints that may be brought before him ; in the latter, there is a resort to magistrates holding office by an independent tenure, as in England, by appointment for life, or by popular election as in this country, which insures proper regard to public opinion.

The prosperity of a country is to be judged of by two tests; 1st, its power of defence against foreign encroachment ; 2d, the protection it affords to its own citizens in the enjoyment of personal immunities and rights of property. The first, when it exists in a large degree, constitutes a military character, and this becomes, in natural course, aggressive, involving a continual state of war. The greatest military power which has ever existed was the Roman ; but the ultimate effect of the martial prowess of that people was, that military service absorbed the entirety of free citizens—private industry was but incidental, and was mainly left to slaves.

In modern states there is far greater regard to the industrial pursuits of civil life. Yet there is a false estimate of the greatness of a nation by simply looking at its military resources. The whole male population of Russia is liable to be called into the army, and no distinction is made whether the levy is for defence or for invasion of another country.

In England, on the other hand, an army is raised only by voluntary enlistment, unless in case of invasion by an enemy, in which contingency all able-bodied men may be enrolled. In most European states conscription is the ordinary mode of providing recruits for the army, and wherever this prevails, there is but little regard to private industry.

It is easier to raise a military force in France,

than in Holland, England, or the United States ; but this does not prove the intrinsic strength of that nation to be superior to that of the others. On the contrary, it is evidence of the impoverishment of the people, when so large a number can be taken suddenly from civil employment into the army. The greatness of a nation is to be measured, not by the number of men that can be readily called into military service, but by the intelligence, wealth, and public spirit of the people. Defence against enemies is, indeed, essential to the independence of a nation ; but this is quite different from aggressive war.

The exclusive pursuit of trade and manufactures has a tendency to impair the warlike character of a people ; yet it has been found, in our own country, that clerks and artisans do very well in military service. Agricultural labor, on the other hand, has been supposed to be favorable as a preparation for the hardships of war, and, so far as respects physical strength and ability to bear the toil of military life, this is, no doubt, true ; but it is said that recruits from cities have been found fully equal to soldiers taken from the country ; this, at least, has been remarked of the recruits sent from Paris into the French army.

In my judgment, the most essential constituency of a great nation is a patriotic spirit. The Roman was taught that his first duty was to his country ;

and so profoundly was this impressed upon the national mind, that Cicero, in the discussion of Ethical principles, gives it a rank above all other virtues.*

The refinements of modern life, while they may, to some degree, enervate physical strength, furnish increased motive for patriotism. As we have larger enjoyment, so have we more to recognize as secured by a national guaranty. But this alone will not suffice unless inculcated as a moral principle. There should be something sacred in the thought of our country, as there is to the Christian in his regard for his religion. It contributes much to this feeling when dangers have been encountered and hardships suffered in contests with other countries, especially when they have been in self-defence. The Scottish nationality was, undoubtedly, made stronger by the ancient and long-continued feud with England.

After a long peace, patriotism is apt to fall off. The prosperity of the United States has, probably, had an effect to diminish patriotic feeling. Our citizens have not been accustomed to think of the protection by which they have been enabled, peacefully and successfully, to prosecute their enterprises. The rebellion of the South, which now agitates

---

*His reasoning is, that the "*Patria*" (country) should be paramount, because it includes all other objects of our attachment. "Dear to us are parents, children, neighbors, friends—but all that is dear in all the relations of life, is embraced in that one bond by which we are held to our common country."—*De Off.*, I., 17.

the country, will not merely be a test of the strength of the Union ; there is to be another effect, which, if I do not much misjudge, will be developed in the future character of our nation. Whatever may be the result of this contest, there will be, at least, a lasting admonition of the value of a united national feeling. It is apparent, that sectional differences of opinion have had more sway over the popular mind, in the Southern States, than regard for the advantages of our nationality. We already see one result of this schism, viz., that foreign powers no longer fear our government, and are taking measures to intervene in the affairs of neighboring states, in a manner which they durst not have done if we had remained a united people.

My opinion is, that the moral strength of the nation will be increased by the ordeal which we are now undergoing. Whether the Southern States shall acquire independence or not, there will be, at least in the North, a vast increase of patriotic feeling—we shall become a more virtuous, and, therefore, a more powerful state.

I predict, that, if the Southern Confederacy should succeed in maintaining itself against the power of the Federal government, it will lack the essential elements of national greatness. We shall become weaker, during a brief period, in our relations with foreign powers, but will soon be able to assert the rights which belong to us as a nation.

We are not understood by European governments. The elasticity and energy which are the peculiar results of our free institutions, cannot be properly estimated where there is no popular independence.

It is my belief that we are to become a greater and happier people, by the effect of the trials in which, by a mysterious Providence, we are involved.

XXIV

SCIENCE OF MEDICINE.

PHYSICIANS—MEDICAL PRACTICE.

It might be inferred that whatever involves the well-being of all classes of men, would be thoroughly considered and understood. Health is so essential to the enjoyment of life, in every possible condition, it would seem that the science which professedly has to do with its conservation, would be reduced to practical aphorisms familiar to every household, somewhat as are the observations respecting farming or cookery.

So far, however, is this from being realized, I know no subject so little understood ; and what is anomalous, there seems to have been less progression in the knowledge of the means of preserving or restoring health, than in any other department of science.

The proof of this will be found in the credence which is given in our own times equally as in an earlier and less enlightened age, to the panaceas or nostrums which are got up by cunning charlatans.

It would be a mere waste of argument if I were
to attempt to expose the devices by which we are
imposed upon. Perhaps a modification should be
made of the general condemnation of popular opinion which I have expressed. It is probable that
many of the medicines which have been advertised
into popular use, have some virtue for certain cures.
But the attempt, which is so common, of getting
up a panacea, or universal remedy, which may have
been of use in some isolated cases, and the success
with which this has been done in a multitude of instances, affords a presumption that there is a superstitious credulity in reference to such matters, surpassed only by the religious bigotry prevalent in
the middle ages. This is shown by the immense
sale of the various medical compounds which have
been brought before the public within the last
twenty-five or thirty years, such as Swaim's Panacea, Morrison's Hygeian Pills, Townsend's Sarsa-
parilla, Brandreth's Pills, and the like. It would
seem as if every man, woman, and child, was suffering from some ailment, if we were to judge from
the enormous amount of such medicines that have
been disposed of.

It does not follow that there is any greater char-
latanry within the period I have mentioned than
has formerly been practised. The instances I have
noted being recent, may better serve for illustration
of the ready credence by which such gross imposi-
tions are encouraged.

I think physicians have not, generally, favored these irregular and unprofessional devices, but their influence has been, to a great extent, neutralized by the inefficacy of their own practice against the course of disease, and especially in respect to those wide-spread and alarming ailments which affect a large part of the community.

It is my purpose to sketch very briefly the general phase of medical practice, with the purpose of demonstrating its inadequacy to furnish the professional aid which is needed. In referring to a very ancient period, I have in view only to show by comparison what advance has been made in our own time.

Æsculapius and his descendants prescribed only for wounds and epidemic diseases. What we call *chronic* ailments—requiring time and regimen for their cure—they refused to treat, or rather they so treated as to dispose of them very summarily.

Wine mixed with flour and grated cheese was administered to Eurypylus at the siege of Troy, for the cure of wounds, by the two sons of Æsculapius. They sucked the wound of Menelaus to extract the poison, and made applications of powerful herbs, but as to what he should eat and drink they made no prescription. The principle was, that if a patient could not get along without so slow a course of cure as *regimen*, he ought not to be cured at all.

Plato says, that this was the general opinion in his own time, and he speaks contemptuously of

Herodicus, who had resorted to medicine and regimen for the aid of his health, although suffering from an incurable disease, and had committed the absurdity of prolonging his life in that way to extreme old age. He remarks, that to persons of a sound constitution, but afflicted with some peculiar diseases, it might be proper to prescribe medicine and resist the disease by drugs or incisions, but not to attempt, by diet, to cure a system thoroughly diseased, so as to afford a long and miserable life to the man himself and his descendants, who would probably be of like habit.

Epimenides, who was called in by Solon to assist in establishing the Athenian democracy, was a physician of much eminence ; but it is related that he made use of sacred mysteries and incantations. There is a fable concerning him, that he slept fifty years, and when he awoke the generation which he knew had passed away. It was after this that he obtained celebrity as a physician.

Medicine was in low repute with the Romans and hardly regarded as a science. Cato, the Censor, wrote a small treatise, in which he prescribes for the diet of the sick, duck, pigeon, or hare, which he considered to be light, though he admitted they had one inconvenience, viz., that of occasioning dreams. He considered fasting of no value, and objected, altogether, to physicians. It appears that they were principally Greeks, and he thought that they

were all under an oath not to use their art for the
benefit of the enemies of their country.*

The most celebrated physician of the thirteenth
century, John De Gaddesden, prescribed, for the
small-pox, that the room should be hung all around
with cloth of a red color, and that the curtain of
the patient's bed should be of the same color.   In
his treatise, entitled " The Medical Rose," he mentions
with approval a treatment for *epilepsy* which
he says, had been used with success.   Mass was to
be said on the feast of *Quatuor temporum*, on Friday ;
and on the Sabbath following, the Scriptures
to be read, containing the account of the disciples
attempting to cast out devils and not being able,
and the text was to be written out and hung about
the patient's neck, " This kind cometh not out but
by prayer and fasting."

In the sixteenth century, *Paracelsus*, in addition
to a discovery, which he alleged he had made, of
a method of transmuting all metals into gold, also
got up a medical prescription for the extension of
human life to any required period. This last discovery
lost popular favor, in consequence of the
death of the Doctor himself, at the age of forty-
eight—it being naturally supposed that he would
have lived longer if he could have done so.

It is not to be wondered at, that physicians should
have resorted to such devices, nor that the popular

---

* Plutarch's " Life of Cato."

mind could be so easily imposed upon by them, when we recall the fact that the circulation of the blood, which is the essential basis of all sound medical practice, was not generally known till the year 1620, when it was promulgated by *Harvey,* an English physician,* and that the functions of the several organs of the human system had not been ascertained by dissections of the body. Nearly all the information possessed by physicians, till a recent period, was founded upon dissection of animals. Doubtless there had been some anatomical examination of the human body, but owing to a superstitious repugnance to what was considered a desecration, no results of such examination were made public ; in fact, it would have involved a physician in popular odium, probably have exposed him to mob violence, had it been known that he had cut up a human body for merely speculative curiosity.

Aristotle, who was the most learned of ancient writers upon natural science, exhausted all the knowledge then existing in respect to animal physiology, but, probably, never dissected the human body.

---

* The circulation of the blood was first demonstrated by Harvey, in the year 1615, and his book, " De Motu Cordis et Sanguinis," was published in 1620 ; yet so little attention did it attract, that Bacon, who lived several years after, viz., till 1626, and wrote some of the most important of his philosophical works in the last five years of his life, takes no notice of it. Bacon's theory was in accordance with the old hypothesis, that the body contained spirits which were pneumatical or gaseous.

Galen, who, until the modern revolution in medical science, had paramount authority with physicians, it is almost certain *had dissected* the human body; this, however, he did n t avow, and the restraint he was under, in this respect, must have diminished very much the illustration he could have given of the constituency of the living human organism. But, it may also be assumed, that his investigation was of comparatively little value, by reason of the necessity he was under of dissecting the human body with secrecy, and, therefore, without aid; and that he was precluded from anatomical analysis upon such a scale as would have enabled him to deduce inferences from the multiform conditions of numerous subjects. Hence, as might be expected, his professional zeal was controlled and counteracted by the difficulties interposed to a serious investigation, which was indispensable to scientific conclusions.

Another embarrassment to the progress of anatomical discoveries, was the want of facility in investigating upon an extensive scale the difference in the conditions of the various organs as affected by disease. This necessarily involved numerous examinations, showing effects of disease upon specific parts of the system. This is now called *morbid anatomy*.

My impression is, that this constitutes the chief element in the advancement of medical science. *The*

most important knowledge a physician can have **is**
that by which he can detect the nature of **disease.**
I have little faith in the efficacy of drugs, and yet,
in ordinary course, it would seem as if the whole
object in calling in a physician was to get an order
upon an apothecary.

Bacon sharply rebuked the practice of the physicians of his time. "Although," he says, "a man
would think, by the daily visitations of physicians,
that there was a pursuance of the cure, yet, let a
man look into their prescripts and ministrations, and
he shall find them but inconsistencies, and everyday's devices, without any settled providence."

Medical writers seem generally to have had a proclivity to visionary speculation rather than to practical deductions. Prior to the discovery of the circulation of the blood, the common division of the
constituents of the human system was into *solids,*
fluids, and *spirits,* the last named being supposed
to be something volatile and easily dissipated.

Bacon was thoroughly conversant with all that
had been written upon human physiology, until
within a recent period (though, as before mentioned,
he seems to have overlooked, or to have contemned,
the discovery of the circulation of the blood), but,
upon the subject above referred to, he says :

"Spirits are nothing else but a natural body rar-
ified to a proportion, and included in the tangible
parts of bodies, as in an integument.   *   *   And

they are never at rest; and from them, and from their motions, principally proceed rarefaction, colliquation, concoction, maturation, putrefaction, vivification, and most of the effects of nature."[*]

The *pulse* he supposed to be a mere sympathy of the spirits between the heart and wrist, similar to what there is between the feet and head.

Sweat he supposed to be put forth by the spirits, and the utility of sweating in certain diseases, as agues, and certain epidemics, was the sending forth of the matter that offendeth.

Stahl, in the latter part of the 17th century, proposed a theory that the *anima*, which was the designation of some unknown power, energized all the animal functions; that it was the elementary principle of life, and that disease was a disturbance of this primary force. A contemporary hypothesis of Hoffman, founded upon observations then recently made upon the office of the *nerves*, substituted *nervous action*, in place of the *anima* of Stahl. The modification was unimportant; it was, in both cases, a mere name of something that was unknown. What constitutes the principle of life was still left unexplained, perhaps it cannot be discovered; but it is obvious that a theory is of little value in which the constituent element is an unknown quantity.

The two discoveries of the *circulation of the blood*, and of the *functions of the nerves*, have been of the

---

* " Bacon's Natural History," pp. 87–88, and 708–711.

very highest consequence in medical science. Another very important result of recent anatomical investigation has been the development of the office of the *glands*. Formerly these had been supposed to be mere *absorbents of moisture*. Equally crude had been the idea of the old physicians, that the *arteries* were *conduits of air*. This was inferred from the fact that after death they were found to have nothing in them, which fact is now accounted for by observation of the return of blood to the heart upon dissolution.

Sydenham was remarkable for keen, original observation of the phenomena of disease, without much regard to theories derived from chemistry, which was at that time crude ; and in this respect he was followed by *Boerhave*. Eminent professors of medicine have, however, still continued to maintain fanciful hypotheses, some of which have had celebrity for a time, but which it would be profitless to notice further.

The most important incident in the practice of medicine during the present century, has been the new procedure introduced by *Hahnemann*, commonly known as *Homœopathy*. The principle upon which this depends is, as expressed by the name, that diseases are cured by remedies that would produce, in a healthy system, *like* symptoms, and is briefly indicated by the apophthegm, *"similia similibus curantur."* The old practice, on the contrary, may be described by the adage, "contraria contrariis curantur." Thus, according to the latter theory, to remove visceral obstruction, a direct force, or antagonism, as of jalap, aloes, croton oil, and the like, must be used. In other cases, with a sort of strategic art, a diversion of the hostile force is attempted by a feigned attack upon some other part of the body, as to counteract inflammation of the *pleura mesentery*, lungs, &c. ; an artificial inflammation is produced externally upon the back, breast, or loins, by *cantharides*, *tartar*, or other irritating substance.

In the homœopathic practice there has been a vast addition to the *materia medica*. Almost every mineral and plant has some virtue in certain cases. I do not think the infinitesimal quantity which is administered by homœopathic physicians is an essential part of the system. It is rather an extravagance, such as is incident to almost every innovation when first introduced.

It is my belief that a great reform is needed in medical practice. The administration of drugs has been excessive, and must ultimately be deemed an abuse. Regimen, including diet and exercise, is of vastly more account for the preservation of health, than all the compounds that can be made by a druggist. In sickness there is a fatuity by which we are led to a superstitious reliance upon prescriptions of strange, nauseous substances—whereas it would be far more rational to trust to the recuperative power of nature, aided by rest, ablution, external applications, and the care of a nurse.

The most successful physicians I have ever known administered but little medicine ; perhaps they had a kind of mesmeric power, and it is certain that in all diseases, except certain forms that are deemed incurable, as *tuberculous consumption, cancer*, and the like, an elastic tone of mind has a singularly sanative effect.

The most interesting, and perhaps it may turn out the most important, inquiry in the whole range of medical science, is the relation of the mind to the body in respect to morbid corporeal functions.

There is sometimes seen an alternation in the course of disease ;—deranged action of the bodily organism is relieved by transfer to the mind, which, in its turn, becomes deranged. Perhaps it would be more proper, in such transition, to consider the nervous system as the locality of the transferred disease. This is the part of our material structure which, so far as we know, is most nearly in contact with the mind ; there may, indeed, intervene some more subtle, ethereal substance, not discernible by our senses, analogous to what has been recently held in respect to the transmission of light, viz., that there is a more rarefied medium than the air, the waves of vibration of which produce the effect which we call vision.

One phenomenon which has been observed in this alternation is, that sometimes when the corporeal system has been relieved of some chronic disease, a change of moral character takes place.*

Incipient insanity is generally indicated by preternatural activity and versatility of mind. There is more ruggedness and force even in argument, but abrupt, and, perhaps, fragmentary and incomplete. The mind is, however, chiefly gifted with quickness of thought and readiness of invention, which passes for wit, but is usually stern, sardonic and repulsive, even while it moves laughter. This was the character of Swift's humor, which was no doubt attributable mainly to latent insanity, that finally broke out into utter madness.

I think it will generally be found that while the struggle is going on with the disturbing force, there is an oscillation between mental and bodily disorder, but when once the mind has yielded to the unfriendly influence, bodily health is improved, that is, so far as respects the functions of digestion, and exemption from pain.

* See ante, "Essay on Health," p. 69.

XXV

DIET.

ETHNICAL PECULIARITIES—COMPARATIVE EFFECTS OF ANIMAL AND VEGETABLE FOOD UPON INDIVIDUAL AND NATIONAL CHARACTER.

THE Romans at an early period lived upon bread and pot herbs, or on pottage.* It is supposed that at one time they had but a single meal in a day. They sat down to it with their servants ; and some of their distinguished men, it is said in the histories of that period, prepared their own dinners.

Although meat was sometimes eaten (as in pottage), their food was chiefly vegetable, and consisting of what was raised by their own hands—the quantity of ground for which purpose was very small, two acres to each citizen being the first allotment—after the expulsion of the kings, seven acres.†

The Turks, in the period of their greatest achieve-

---

* Cheese and eggs, or perhaps meat, sodden with garden vegetables ; it is no entirely clear. See " Adams' Antiquities," 471.

† This allotment was called " hæredium," or " sors." Many of them might be acquired by a single citizen, but one was considered sufficient. *Cincinnatus*, *Dentatus*, *Fabricius*, &c., had no more ; Cincinnatus had but *four*.

ments, as is related by a traveller, fed upon bread, garlic, and sour milk.*

In Persia it seems that the ordinary fare is chiefly vegetable. Cold boiled rice, bread, and sour curds, are the provisions usually carried upon a journey.†

The Prussians subsist chiefly upon bread and butter, and potatoes. The lower class of Irish, it is well known, live almost entirely upon potatoes. The physical strength which they possess, and their capacity for labor, prove that their diet is, in that climate, not inadequate to their physical needs.

In the northern regions, animal food seems to be craved. The Esquimaux eat enormously of the fat meat of the walrus, and drink whale oil. Parry says, that a single person will consume ten or twelve pounds of solid food and a gallon of whale oil in a day.

The Siberians also consume an incredible amount of animal food; but in the severe cold of the region which they inhabit, human beings are reduced almost to a level with the brutes.

· Meat diet is more used in England, and especially in London, than in any part of Europe. The annual consumption of meat in London has been estimated at 143 lbs. for each individual; but as there is a large number who get no such amount, by reason of their poverty, the proportion is, in reality,

---

* " Busbequius' Travels," quoted by Sharon Turner.
† Morier's " Haja Baba."—*Id.*

much greater for those who can indulge a fleshly appetite. In the whole of Great Britain, the average is 92 lbs. for each individual ; in France, 36 lbs., though in Paris it rises to 86 lbs.*

In our own country, animal food is the chief article of diet ; and its excessive use has induced, as is now generally understood, the national complaint of Dyspepsia or Indigestion. Other causes have doubtless contributed, as intemperance in the use of alcoholic liquors, which we have inherited from our English ancestry—but more especially the great cerebral activity which has been developed in our people, and which must have had a direct tendency to overtask and derange the nervous system.

The tenacity with which certain modes of diet are adhered to, during long periods, in the same locality, is an interesting subject of speculation. The natural productions of a country, of course, must have had chief influence in determining the original habit, but in most parts of the world there is sufficient variety of products to admit of some choice, and so far a national habit may be deemed arbitrary. Again, assuming some article of diet to have a large preponderance in nutritive or agreeable qualities that would account for its general use, still there

---

* "Turner's Sac. His.," v. 3, 349. There is probably no great accuracy in such calculations, but the general fact is sufficiently established that there is far greater consumption of animal food in England than in any European country. This cannot be the effect of climate, as the Irish do very well without it.

may be observed a singular uniformity in the mode of preparing it, which, without other cause than the law of custom, will prevail for centuries. Rice is the common food in most Asiatic countries, but the Tartar races have, from the earliest period of which we have any historic tradition, prepared it with meat in the same way that is now done—being the well-known *pillaw*.

The Russian *schtshi* (*cabbage-soup*) seems to have obtained equal universality, though, perhaps, not of equal antiquity. It has become an article of diet, throughout the empire ; of the aristocracy as well as of the peasant ; to the latter it is, indeed, almost the sole sustenance of life.

Some observations by Kohl, the German traveller, upon this subject, are worth transcribing :

" *Pillaw*, the well-known tower of boiled rice and pieces of mutton, which still occupies the centre of the dining-table throughout the whole of the East, smoked on the boards of the ancient Persians and Parthians in the times of the Greeks and Romans, and there is no doubt that many a Babel tower of stone and marble will rise and sink before that tower of rice, which rises afresh every day, shall be destroyed."[*]

He also mentions the ancient manufacture of sau-

---

[*] " Kohl's Russia," c. xxii. The *schtshi* he describes as composed of white cabbages, barley-flour, small pieces of mutton, and *kwas* (which I understand to be a home-made beer). The poor, however, omit the meat.

sages, at Byzantium, which, he says, is still continued, with little variation, at Constantinople.

The proscription of any article of food is of equal
importance. The aversion to swine's flesh by the
Arabians and kindred races, including the Jews, is
not, as yet, sufficiently accounted for. The laws of
Moses might furnish an explanation in respect to
the Israelites, yet, unless there had been some national adaptedness thereto, the laws would hardly
have been enforced.

As to the effect of animal food upon the human
system, it is ascertained by the observation of many
persons in various countries, not to be necessary for
a sound physical condition ; in fact, it appears to
be demonstrated, that it is, on the whole, not so
favorable to health, or long life, as a vegetable diet.*
It may create greater muscular activity and physically affect the mind in the same way, that is to say,
may give it greater energy ; but this, I think, is rather
to be designated as something akin to mere animal impulse. That species of mental vigor which is usually
conjoined with a high degree of corporeal functions,
belongs to the physical, rather than the intellectual
part of our system ; clearness of mind, sagacious perception of truth, and the sense of right, belong to an

---

* This remark may be subject to some modification as to those parts of the world
in the extreme northern latitude, as in Greenland and Siberia, where the climate
seems to induce a craving for the grossest kind of animal food—as of the walrus
and whale. Yet in Lapland the diet is chiefly the milk of the reindeer. (See
Brooke's " Winter in Lapland.")

organism whose physical impulses are not stimulated by much animal food, but whose natural wants are merely satisfied and no more. The low, or vegetable diet, has been found most consonant with a healthful condition of mind, and it seems, also, best to insure exemption from bodily disease.

Abstinence is well known to be an important remedy for the cure of most ailments, particularly any tendency to diseases of the head, heart, or arteries. The fasts which were enjoined by the Church in an early period, may have had their origin in some considerations of the effect upon the bodily system. The most zealous Christians have been always addicted to a spare diet. The monastic orders were uniformly, in their inception, restricted to a vegetable diet ; even to a late period it was the prescribed rule, although sensual appetite became, in time, in almost every monastery, more powerful than ascetic devotion.

Many men distinguished for great intellectual labor, have been abstinent to an extreme degree. The instances are too numerous to leave a doubt that such a mode of living comports with the highest exercise of thought.°

It has been said that men become ferocious in proportion as they are carnivorous. In civilized

---

* *Winkelman* lived upon bread and water ; *Dr. Adams* (the author of Roman Antiquities) lived chiefly upon oatmeal porridge, and this is said to have been the diet of many contemporary Scotch scholars. (See " Turner's Sac. His.," iii. 341.)

life, I should say that animal food has a tendency to develop gross propensities, not necessarily ferocity ; it may be as well sensuality ;—chastity, and almost every other virtue, are best maintained by a sparing diet ;—the vegetable, being the most favorable. The opinion of Lord Byron, that the reason women were better in disposition than men was that they did not gormandize as much, was, perhaps, well founded, though it is to be observed that the greater delicacy of the female structure and bodily functions naturally induces, or is necessarily associated with, a corresponding refinement of mind and aversion to gross habits. Women who are subject to labor in the fields, however, eat as heartily as men.

XXVI

POPULATION.

THE principles upon which population depends are, as yet, imperfectly known. The facts upon which any reliance may be placed are of recent date, and sufficient time has hardly elapsed to furnish the means of establishing a comprehensive system. Ancient statistics are meagre and entitled to little credit.

Thus, Diodorus Siculus says, that the city of Sybaris had 300,000 citizens able to bear arms, and that that number actually encountered in battle 100,000 citizens of Crotona, another Greek city— yet Sybaris was not a trading town, and had only the advantage of fertile valleys around it, favorable to agriculture.*

The same author states, that Agrigentum, when it was destroyed by the Carthaginians, had a popu-

---

* See "Hume's Essay on the Populousness of Ancient Nations." I am indebted to this erudite essay for many of my criticisms upon ancient authors.

lation of 20,000 citizens and of 200,000 strangers, besides slaves and women and children, which would make an aggregate of more than 1,000,000 ;* Diogenes Læertius says 800,000. Yet the whole resources of this city consisted of a small district of country, fertile in wine and oil, which were exported to Africa.

Diodorus Siculus ascribes to Egypt a population of 3,000,000, which was probably below the real amount, but says that there were 18,000 cities, which is absurd.

Polybius says, that the Romans and their allies mustered 700,000 men, able to bear arms, between the first and second Punic wars. Diodorus Siculus says 1,000,000—but this is a greater number than the same extent of country (the Pope's dominions, Tuscany, and part of Naples) would now produce.

It is related that Dionysius, the elder, had a standing army of more than 100,000 men, and a fleet of 400 galleys ; but Sicily was an agricultural country, and though the inhabitants might have been numerous, it is impossible that it could have supported any such body of mercenary soldiers.

The population of *Athens* is stated by Athenæus* to have been, according to an enumeration of Demetrius Phalerius, 21,000 citizens, 10,000 stran-

---

* Hume estimates it at 2,000,000, **but this is too large.**

10*

gers, and 400,000 slaves.* The proportion of slaves is out of all bounds of probability, and it impairs the credibility of the whole statement.

According to Demosthenes, there were 20,000 free citizens. Thucydides stated the military force at 13,000, which, of course, included men of full age—calling these one fourth of the inhabitants, it would make, with women and children, from 50,000 to 80,000—to which add strangers and slaves, it might make 160,000 to 200,000. Xenophon says, that there were but 10,000 houses in Athens, which would make the population less than the above estimate. Again, the census of taxable property in Athens, is stated by Demosthenes at 6,000 talents —but the lowest price of a day's labor of a slave was an obolus, which, if there had been but 40,000, would have made alone a much larger amount, and this would be a sufficient proof of the extravagance of the statement, that there were 400,000 slaves. The walls of Athens, it is true, were, according to Thucydides, eighteen miles in extent, besides the sea-coast—but Xenophon says that a great deal of waste ground was included.

The population of ancient Rome. cannot now be determined with any certainty—even the extent of the city is the subject of controversy among critics,

* As this means only men of full age, capable of bearing arms, the result would be, that including women and children, the entire population was about two millions.

some estimating the circumference at **thirteen miles, others at thirty.** The number of houses **in the** reign **of** the first Theodosius was 48,382, **which,** allowing twenty-five **persons to each house, would give a** population **of** 1,200,000.† But this **is far** below what is commonly supposed.

If there is such imperfect knowledge of the number of inhabitants in **the** two most celebrated cities of antiquity, it is obvious how utterly **unreliable** must be all estimates of **the** population **of an entire** country, especially **at any period of** time **before** statistics **had become a matter of** public **interest.**

Some general **observations upon the natural increase and** decline of population, **and many authentic facts of recent** date, lead to an inevitable conclusion, that the number of inhabitants in any **well-**governed country **is** greater at the present time **than**

---

* Hume concludes, from a **passage** in **Pliny, that the average length was about** five miles, and breadth two and **a half miles.** Gibbon estimates the circuit of the walls, at the time of the siege of **the Goths, at twenty-one miles.**—2 *Gib.* c. 31.

† The number of **twenty-five to a house might** seem to be very large, but is deduced by Gibbon from a **comparison of the dwelling-houses in** Rome and Paris. It appears that the plebeian habitations in Rome were of great height, and such tendency was there to **a dangerous excess, in this respect,** that imperial laws were passed restricting **the** height to seventy feet. This would admit of six stories, **of** moderate elevation, and, by **taking the** number of inmates in a house of corresponding dimensions in Paris, at this day, it is made probable that the estimate above stated is too low. **There is a** circumstance **which** would much contribute **to** this result: modern civilization has largely **added to the indoor comforts of life** and this induces something of exclusiveness in the occupation **of premises, even** it be but a single room ; but the common people of Rome **were, for the most part** in the street, the forum, **bath,** or amphitheatre ; **and went home only to lodge, and** as may be supposed, a single lodging apartment would accommodate many persons with such rude provision as was then deemed sufficient.

at any former period, and that the aggregate, in the civilized part of the world, greatly exceeds the population existing within the same limits under any of the ancient forms of nationality.

I. Population depends mainly upon the means of sustenance, that is to say, it would be so when other drawbacks, such as oppressive tyranny, do not intervene, and, therefore, in the ordinary course, population is increased in proportion to the increased productiveness of the soil, and this, again, depends upon the state of society, and the encouragement offered to labor.

A merely agricultural country does not sustain so great a population as one in which trade and manufactures flourish.*

An easy subsistence, that is, the enjoyment of all the necessaries of life, conduces to marriages—but still more to the bringing up of large families, as the result of marriage. Poor people will marry, even when they have not the means of support, but they rear a less number of children than those who are in comfortable circumstances, that is, when the latter are not subject to counter influences, as from vicious or enervating habits. Smith has observed, that soldiers have as many children as other classes

---

* Smith held the opinion, that countries are populous in proportion to the *food* produced, without regard to other necessaries, as clothing and the like.—*Wealth of Nations.*

But the true rule undoubtedly is, that it is in proportion to *all necessaries,* of which food is, of course, most important.

of men, but that a larger proportion of their children die in consequence of hardship, or want of proper care.

On the other hand, it is a remarkable fact, that the Irish peasantry, of the poorest condition, rear large families.   How much is to be attributed to climate is not settled—perhaps the vegetable diet, which is general, may conduce to health, and, therefore, to longevity.   I have observed that large numbers of the children of Irish who have emigrated to this country, die at an early age.   There is, indeed, greater mortality in every period of life, among these emigrants, than among our native born citizens who are in a fair condition of life.   This difference may, to some extent, be accounted for by the greater prevalence of intemperance and uncleanly habits among the Irish.

II. We may observe in small towns where there is some trade or manufacturing, and a quick market for agricultural products, the lands of the adjoining country are improved—the people are generally industrious—marriages take place early, and large families are brought up.   But when there is little or no trade, lands are likely to be poorly cultivated—enterprise will be deficient, and the young people may, in general, be expected to be inert and marriages to be infrequent.   This may be seen even in our own country, though there is here a general growth in wealth and population, yet in some parts, where the influences tending to produce this growth are interfered with, as by emigration, change in the course of trade, and the like, a shiftless habit will be found to be the prevailing phase.   In the old countries, where the causes which have operated so strongly to stimulate enterprise here, have, comparatively, slight effect, it may be well conceived how population is retarded by defective industry and want of trade.

Where, however, lands are divided into small farms, the production will be larger, and so, consequently, population—an example of which we have in the proprietorship of lands by the Romans, at an early period.   Ownership of the lands by the farmer is most advantageous to a people—but this has been the rule in but few countries.   Large estates are usually held by a few—the Romans employed slaves —in most European countries leases are made, and the length of these and the amount of rent charged determine the condition of agriculture.

France was impoverished before the revolution of 1789, by the heavy burdens upon the tenants of lands.   The breaking up of the large estates and the distribution of lands among small proprietors, was one of the greatest benefits of the revolution—the salutary effects of which was demonstrated by the fact, that, notwithstanding the destructive wars carried on from the commencement of the revolution till 1815, the population of France had increased, at the census of 1817, to 29,000.* It has since then increased to considerably upwards of 30,000,000— but the advance since the termination of the war is mainly attributable to the change in the condition of the agricultural population.

III. Security of property and private rights is essential to the growth of population, and, in modern times, the superiority in this respect is so great over former periods, that it is an argument almost undeniable, that popnlation must be greater now than formerly.

The advantage enjoyed in this country is, mainly, that every man, however humble, is entitled to the protection of the laws, and can enforce it. On the other hand, under despotical governments, the administration of laws is partial. The peasant has little chance in a contest with a wealthy landholder.

IV. A last consideration affecting population is a sound moral state of society. Vice destroys industry, and by consequence the means of support, and, therefore, interferes with marriages and a proper condition of domestic life.

I think the main circumstance which has been efficient of the vast growth of population in this country is, that there has been such remunerative

---

* At the commencement of the revolution it was a little upwards of 27,000,000.

return for enterprise that early marriages have been safe, and that there has been such inducement for industrial exertion, that no class of our community has been allowed to remain in a state of idleness. In other words, industry has been the general phase, and, by consequence, a comparative condition of domestic purity, and it has not been respectable to live in idleness, even when a man had the means of indulging in such a propensity.

V. Some interesting statistics of a recent date will be pertinent to the inquiry which is the subject of examination.

The densest population in Europe is in Belgium, which has 3,791,000 inhabitants to 13,000 square miles, being an average of 290 to the square mile, but in some parts of the country amounting to 500. —and this population is mostly agricultural, and not much congregated in cities.

Holland has 2,745,000 to about 11,000 square miles, or about 247 to the square mile—but in Holland there is great facility of intercourse between all parts of the country by water, which, of course, makes a ready market for agricultural productions, and induces a high state of cultivation ; and again, the habits of the people have, during several centuries, been industrious and economical, and the civil government has, on the whole, been the best in Europe, as far as respects internal prosperity. In the seventeenth century the foreign commerce of

Holland was greater than that of all the rest of Europe.

England had a population, in 1377, according to the result of a poll-tax, of 2,300,000,* but this is supposed to be much under-estimated. In 1575 (reign of Elizabeth), the census made the inhabitants 4,500,000. In 1801 the population was 8,331,434. In 1831, upwards of 13,000,000.

It is said that one half of the population of England is in cities.† This population is principally employed in manufactures and foreign trade. Agriculture has, however, improved with the increase of population—the production of animal food is greater than in any other country, in proportion to the number of inhabitants—one half of the land, it is said, is devoted to pasturage—grain is raised in sufficient quantity for the average consumption of the people, but occasionally, as is the case during the present year, a large quantity is imported from other countries.

---

* In the reign of Edward III.
† See article on British Population.—*Museum*, 1828, vol. 1.

XXVII

PROBATION OF LIFE.

PAIN OF BODY—ITS PROBABLE USES.

THE heathen believed all human suffering an in-fliction by deities hostile to man, and they sought to avert their cruelty by such offerings as would be adapted to appease human passion. But there was no idea of moral discipline, or, at least, of divine goodness, in the ordering of the affairs of life. We are taught, on the contrary, in the Scriptures, that all human suffering is a chastisement of sin, not vindictive, but for an ultimate good.

If suffering be regarded as a discipline, another question painfully presses upon us, whether the amount of apparent good obtained, bears a just pro-portion to the pain endured. This is involved in perplexity, so far as the solution depends on mere human reason. The following observations will tend to put before us something of the difficulty in measuring the proportions of pain, and its salutary consequence, and, perhaps, suggest some explica-tion of that difficulty. *First.* As to the extent of pain which the human organism can endure. The suffering which we have to undergo in some acute diseases, naturally excites appalling apprehensions of what we may be subject to in the final struggle, when life shall be extinguished by the forces of destruction. Yet, the closing scene of life is often, in fact, peaceful, and apparantly free from pain. To my mind, it is a more fearful thought how much can be endured before the vital power is overcome. Under the intense torture of some forms of suffering, I sometimes feel a shrinking back, a dread, almost despair, in the reflection how much more can be sustained before relief shall come by the parting of the soul from its corporeal organism, and the cessation of all sensibility. But here intervenes a provision which seems intended as a merciful interposition in aid of suffering humanity, that pain, beyond a certain degree, ends in unconsciousness, at least what so appears. Whether there still remains sensibility to pain when the external expression ceases, cannot be known, as there remains no recollection afterward of what has passed while the sense of external things has been suspended. A disturbing apprehension might indeed arise from what we do recollect of sensations in a semi-conscious state ; when the perception of things about us has diminished, that external anguish has been none the less acute. But this, probably, is brief in duration, though in the actual endurance it may seem long protracted. Such is the illusion of frightful dreams.

It is supposed that what we recollect has passed through the mind, in the transition from sleeping to waking, perhaps in a single moment, though in the retrospect the time seems long.   On the other hand, it has been said that the extent of pain in a mortal disease is not in proportion to the pain sometimes suffered without endangering life, as Bacon* has remarked, that "death sometimes passes with less pain than the *torture* of a limb," for which he gives as a reason " that the most vital parts are not the quickest of sense."   But this, as a general rule, is not well founded.   It is reasonable to suppose that pain is in proportion to the resistance of the vital powers to the principle of dissolution.   Therefore, in acute diseases, which destroy life suddenly, as congestive fevers, inflammations of the lungs, viscera, or other vital parts, there is great suffering.   There is, however, a point where sensibility apparently ceases.   In Asiatic cholera, the countenance indicates distress, when sometimes the patient will say he feels no pain.   Here the disease seems immediately to overcome all resistance, while in common · cholera, which is less rapid in its progress, there is extreme suffering.

I may here notice a theory, sustained by some remarkable proofs, that bodily pain, when excessive, terminates in a pleasurable sensation, such as what is related of a young confessor put to torture, who

---

* Essay on Death.

said that after the first agony he was refreshed and even exhilarated.* So in the case of persons resuscitated after life was nearly extinct by drowning, it is related that after the first paroxysm of suffocation, a delightful languor succeeded, and pleasing images of scenes far distant, and events of early life, recurred to the mind. A similar effect has been spoken of as recurring after hanging, where there has been restoration of life. These incidents, however, though they may tend to relieve our apprehension of suffering in death, do by no means disprove the actual extent of pain which can be, and is endured without extinction of life. Few persons have been without experience, at some period of life, of intense anguish from acute disease or bodily injury, and all, perhaps, have had the opportunity to observe the sufferings of others, which were unmistakable in their poignancy. Perhaps there is a kindly provision which sometimes affords relief, and in most cases of severe distress, perhaps there is more in appearance than is actually endured, yet the general fact remains still true, that there is a fearful extent of human suffering.

Secondly. There seems, to common observation, to be little correspondence in the extent of suffering to age, sex, or any condition of life. Does the

---

* See "*Moore's Relation of Body and Mind*," p. 287. He also mentions a case of a traveller in Africa, who nearly perished with thirst, but in his sleep found relief in the most delicious sensations.

man of robust frame and fully-developed organs suffer more in sickness than the little child ? We look upon the latter with peculiar sympathy, by reason of its helplessness to resist whatever oppresses it. If my recollections of early life are reliable, I suffered greater anguish then than I have in maturer years, or, at least, had a greater sense of it. After all, it depends on sensibility, and it may be that in infancy it is greatest. Females of delicate organism exhibit it in greater degree than men. Do they suffer pain in sickness and death in the same proportion ? Perhaps excessive sensibility is sooner relieved by the exhaustion of nervous energy. But it is fearful to think of the sufferings of fragile women in sudden disaster, as by fire, or wreck at sea, or the more appalling form of death by brutal violence.

Third. If there be no religious principle, it is usually seen that pain at first rouses a bitter feeling, as if there had been some harsh and cruel treatment. The heart which has had no hallowing influence of pious emotion, intrenches itself in a stern resolution—a defiance of what it deems a pitiless arbiter of human life, and if the suffering be not long protracted, such defiant resistance may last to the end ; but if the pain should be lingering, such resolution will be likely to subside into abject superstition. The priestly office will be invoked—some clergyman will be called to stand between the soul and its God ; **but** it will still be **the same feeling** that actuated the heathen worshipper—servile **prostration** before a merciless Deity—but **no contrition**— no consciousness **of a deserved** penalty—no sense **of Divine mercy mingled with the** chastisement of sin.

On the other hand, **the soul that has a consciousness of** its **natural** state, **and** has obtained a clear perception of the holiness which is the attribute of God, will feel a conviction that there is necessarily a penalty **for** transgression, yet, that in **the punishment** there **is a purpose of mercy.**

Fourth. A physical advantage, **or rather a prophylactic principle, is connected with** pain **of body. It is** a **perpetual admonition against the** violation of **the laws upon which** bodily **health** depends. The **child very** soon learns to avoid what is destructive **to its well** being by **the** severe lesson of pain. **A** similar warning is met with in youth against **the** indulgence of unlawful **passion. In** this **case, it is** true, **the consequence is not so immediate, at least the entire injury that will follow by persistence in a wrong course does not ordinarily ensue at** once. **The evil** consequence **does, indeed,** begin as soon as **there is** transgression, **and in** some instances is, in its very inception, irreparable. Usually, however, there is **opportunity for repentance and reform. With** the same **certainty that fire will destroy or** impair the fleshly organism, **will vice sooner or later entail** disease and suffering **upon** the bodily system.

There seems to be a moral discipline intended by the slow progression of this physical penalty. A longer impunity is permitted to one than to another. In no instance is the exact time and manner of the retribution calculable. Delay lulls present fear, and the very fact that one outlasts another in the same guilty course, is apt to induce a vain opinion that there *may be entire exemption* from evil consequence in a particular case. Common observation ought, indeed, to be sufficient to satisfy any one that retribution, at some time or other, in some form, is sure; at least it may be inferred from what we see of the effect of flagrant vice, that all lesser degrees have also their measure of penalty, although not so strikingly obvious.

"Thou makest me to possess the iniquities of my youth," said Job, and this may apply to bodily suffering as well as to anguish of mind. Indeed the former often constitutes the memento by which early transgressions are brought in array before the mind in after-life. In the prayer of the psalmist, "Remember not the sins of my youth," we may suppose that the disease of which he was suffering recalled to his mind early transgressions. One thing is certain, that according to the common testimony of men who have been profligate in their lives, there is an aggravation of bodily suffering when the constitution is at last broken. Infirmity must, indeed, come to all, and no one is exempt from pain ; but there are pangs in bodily suffering, caused by vice, oftentimes as marked as the distress of mind which is the consequence of remorse. Perhaps it is safe to say, that every misconduct in life will rise up again in judgment. How often do we see men overtaken by the consequences of transgressions committed long before, so that it would seem as if they had been pursued by an avenging spirit. It becomes, sooner or later, the conviction of those who have been depraved, especially when they have been guilty of crimes denounced by human laws, that they are hunted by their sins, so that their minds become subject to superstitious terrors.

Fifth. But while we thus see that there is a retributive consequence of all misconduct, yet it cannot be maintained that physical suffering is in exact correspondence therewith. There are, indeed, many other modes of retribution. And again, we have before us instances of disproportionate suffering of those whose lives have been comparatively innocent. In some of these the cause may be hereditary predisposition, but in others, it is accidental and inexplicable. There is, however, this counterbalance, that when the character is virtuous, there is generally alleviation for distress by the sympathy of others. The innocence of childhood appeals powerfully to the kindly emotions of our nature. Even the vile and hardened are not insensible to it. With what interest is a suffering child watched over by its

11

natural protectors ! So the good and amiable, in every period of life, attract to their aid, when subject to distress, the kindness of all around them. The greatest solace, however, to suffering humanity, is religious faith. Not only has this been found sufficient to sustain the soul in the ordinary pains of life, but under all the inflictions of human cruelty in Christian martyrdom.

It is sufficient to mention this. It is not my present purpose to illustrate, at large, what may be gained from a genuine experience of evangelical faith. There is, indeed, one result from this discussion which cannot be passed over, viz., that in the arrangement of human affairs by a Divine Providence, the lapse of time, according to human modes of calculation, is of little account. The time which has passed may be infinitesimal, in comparison with the great future of human existence upon this earth. It is true the apostles supposed the end of the world near at hand—but a long progression has since taken place, and we are now warranted in supposing that it will go on until a renovation of the entire human race shall have been accomplished.

I may say here that an error of the first disciples, in respect to the termination of the system to which human life is subject, does not involve the assumption that they were, therefore, mistaken in the principles by which human responsibility is to be determined.

We are rather led to the conclusion, which is in analogy with all the history of the human race, that the course of renovation is progressive, and that a long period may elapse before the character of men shall be brought up to the standard prescribed by our Saviour—nor, perhaps, is it possible that it shall ever attain to it. It may be that a general approximation is all that is to be hoped.

XXVIII

PROBATION OF LIFE.

SUCCESSIVE CHANGES—OLD AGE—DEATH.

———

THE periodical change in the progress of life involves a principle somewhat different from the trial incident to every part of life, from pain of body. When no accident intervenes, there is a regular progression from infancy to old age. Every part of this is indeed subject to pain, disease, and death. It is the exception, rather than the general rule, that any one shall reach to old age. Accidents and acute diseases terminate the lives of the larger number before arriving at old age. Hence, *Montaigne* insisted that it was as natural to die early as late, and by a violent death, as by the infirmity of old age ; nay, even more so, because the majority of mankind do not live beyond middle life. Cicero, on the other hand, denied that Nature could be supposed to have neglected her own work, and comparing human life to a drama, that it could not be supposed the fifth act would be neglected. Old age he therefore considered to be the natural consummation of life, "*tanquam* **arborum** *baccis terræque frugibus maturitate* **tempestiva quasi** *vietum et ca-*

ducum" (like to the production of trees and the fruit of the earth, bending under the weight of autumnal maturity.)

It is indeed true, that there is an apparent order of nature, which is developed in the successive periods of human life, old age being the consummation of a completed course. But, when we consider how few of all that have lived attained to the prescribed limit which, according to this theory, is necessary for the accomplishment of the purpose of life, it becomes obvious that, as respects the largest part of our race, it was not designed that they should live through any particular period, which should have in itself a completeness for any specific object. On the contrary, life is determined at every stage of existence—most frequently in infancy and childhood. All that can be said, therefore, is, that in the comparatively small number of cases in which old age is arrived at, there is a more perfect course of life.

To that period, when it is attained, the following observations will apply : Old age is not necessarily a state of imbecility ; the vices of youth may, indeed, produce a wreck of the mind as well as decrepitude of the body, but the weakness of old age, which is not affected by such causes, is in itself respectable. *Senile garrulity* is indeed proverbial as something belonging to old age, but a proper exercise of our faculties will prevent any disgusting display of worn-out nature. " *Arma senectutis artes*

enercitationes que virtutum" (the arms for the defence of old age are exercises of the mind in the virtues of life, and the rational employment of our
faculties.)   A virtuous life and proper employment
of the mind, will insure that happy old age which
is like the frugiferous autumn.

Disease, however, is incident to every stage of life,
and whenever it occurs, whether early or late, it is
genitive of the same discipline.   The same principles may, therefore, be considered as applicable to
every stage of life, when infirmity of body intervenes.
One general remark is all I have to add to what has
been already sufficiently developed, viz., that there
is a mysterious provision in the constitution of our
nature, by which pain, decline of health, and the
process of extinction of life, in whatever period it
may occur, has its alleviations—not derived from
speculations of philosophers, nor from any resources
of recondite learning.   An uneducated man has in
himself a secret power of bearing pain, and of looking calmly upon the dissolution of his corporeal
frame, equal to what Seneca attained by all his
erudite meditations.

Montaigne has noticed, with his usual discrimination, how little difference there is between the
peasant and the scholar in the capacity to endure
the trials of life, and hence deduces the fact that
philosophy (meaning by that the speculations of
learned men) is of very little practical use.

We now come to the consideration of *death*
The greatest trial of life is the thought, which is
continually forced upon us, that we must die.

Is it not mysterious, nay, one of the greatest
problems of life, **that all** men fear death ? **though,
as Bacon says,** "*it is as natural to die* as to be
born." This fear, in most men, is perhaps not
overdrawn by the great poet :

> The weariest and most loathed worldly life
> That age, ache, penury, and imprisonment,
> Can lay on nature, is a paradise
> To what we fear of death,
> SHAKESPEARE.—*Measure for Measure.*

It is not easy to decide whether the apprehension
of the pain and shock of dying, or a revulsion at
the untried and unknown future which lies beyond,
has most effect to oppress the mind.

There is, doubtless, exaggeration of the suffering
endured in the pangs of death ; but there is, **per-**
haps, greater dread of the mysterious change of existence which is to take place, **of which our** experience furnishes no analogy. We **do, indeed,** have a
partial suspension of the functions **of life in** sleep
—yet **vitality is not** suspended—the pulsation of
the heart, the regular inflation and compression **of
the** lungs, the warmth and softness of our fleshy
fibre, voluntary motion not wholly intermitted, **and,**
finally, a consciousness of an easy play of the mind,
as if acted upon by exterior sensation—these take
from what is called the image of death **all** that ap-

pal us in the cold, unbreathing, and stiffened form which death presents to our view.

I believe (although it is but a hypothesis) that our dread of death is chiefly owing to the circumstances of contrast with what belong to, and are essential to life. Dissolution is preceded by the loss of the control which the mind has had over its corporeal organism—it is consummated by the extinction of all the outward signs of sensation. Yet, still we cannot, in looking at the lifeless body, put away the habitual association by which the invisible life is indissolubly joined, in our thought, with the exterior form and features which have been the visible expression of whatever constituted the individual. Hence we still look upon those lineaments which have, to our eyes, represented the living being, rigid though they have now become, and cannot separate in idea the life from the form which it animated. We cannot conceive of annihilation—nor can we wholly realize disjunction of life from that which was the living, sentient, and acting corporeal form. Unconsciously perhaps, at least without any distinct analysis of the thought, we still attribute some sort of sensation to the inanimate body. We involuntarily shudder at committing to the cold ground the delicate form, once so carefully guarded from exposure to the inclemency of winter and the damp evening air—and we feel oppressed in thinking of the close coffin, and of the compact earth piled above it.

Reason as we will, such thoughts will rush upon **the mind** when **we** think of death. **It** was not mere poetical invention **that** assumed a sort of sentient existence as belonging to the dead. The *Egyptian* believed it. **The Jewish** prophets, though divinely taught, still **used** language and imagery that prove how deeply they had been impressed by a similar idea. (See Isai., 14 ; Ezekiel, 32.)

In the contemplation of death, those who are in the vigor of health have more fear than those who are feeble or diseased. The reason is, that the contemplation is **familiar** to the latter, while it is forced **upon the former only** upon extraordinary occasions ; and, again, the contrast is more striking between **the** full enjoyment of all our powers, and the state **of** inanition in death, than there is when we already feel the loss of a considerable part of what has constituted our life. But it is also true, that in health we do not often think of death at all.

There is one consideration which should have a great influence in reconciling us **to the** thought of death, and **imparting** to us equanimity when we find it approaching. It is that the weak and timid, the delicate and sensitive woman, the child gently guarded by parental care, have to pass through whatever we fear in death, and that the fragile and helpless, as well as the strong in heart, must **enter** into the gloom and mystery which envelop the scenes of the invisible world—and not merely in the midst of friends, **and** with all the solaces of affectionate

11*

care, but how often in the midst of the terrors of
violence, in the dark tempest and raging ocean, or
in the fury of conflagration, or in the more fearful
scenes of human outrage ?

Shall we shrink back from death in its ordinary
forms, when it has been endured by countless numbers of our fellow beings under the aggravation of
all that is terrible to think of ?

Recall the memory of the loved ones who have
gone before us in this last trial of human life ; the
sister, lovely in form and angelic in her disposition ;
the child, trustful in its parents' love ; the wife,
mother, friend—beings whose affections were inter-
twined with our own—who in life were, perhaps,
girded by our sympathy and sustained by our
stronger resolution, yet parted from us to enter
alone upon the experience of what is so appalling in
thought, even to the strong-minded.

It will be an appropriate conclusion of this essay
to recur to that surest solace of the soul in all the
trials of life, viz., a true Christian faith.   It would
be a mistake to suppose that in this is to be found
a solution of all the mysteries that environ us.   It
consists, not of speculative knowledge, but of trust
in God.   "Thou wilt keep him in perfect peace
whose mind is stayed on thee" (Isai., xxvi, 3), said
the devout prophet.   But there are degrees of faith.
We may believe that all things are ordered by a wise
Providence—but it is another thing to feel a personal affinity to God—to be in communion with

Him, and to have the consciousness of his paternal goodness extended to us individually. The soul, imperfectly enlightened by divine truth, may still be subject to distressing doubts respecting many things, the solution of which has always baffled mere human reason, and to fears as to the future, such as disturb the unrenewed mind.

But against all such fears, and against all the pangs of disappointed hope, against every form of suffering incident to human life, the soul that is at peace with God is fully armed.

If you are, indeed, a child of God, and so far as human imperfection allows, conformed to the divine will, what cause is there for fear, since the elements of nature and the course of human life are all subject to that will? The turbulence of all the forces of nature, the tempest, the flood, famine, and pestilence, are swayed by the power of the Omnipotent.

So, also, the malevolence of wicked men, why should it be feared, when we know that the world is under the government of a righteous Judge?

" Surely the wrath of man shall praise thee, the remainder of wrath shalt thou restrain" (Psalms, lxxvi, 7), and if there be any invisible powers of evil more to be feared than the wickedness of man, even these are subject to a higher power, by whose aid we may successfully "wrestle against principalities, against powers, against the rulers of the darkness of this world" (Eph., vi, 12.)

Therefore, **the humblest disciple of Christ,** if he have that measure **of** faith which **we have supposed,** will have rest **of** soul even if he cannot "**understand all** mysteries,"—nay, he seeks not **to** understand **them,** but reposes with child-like confidence upon **Him** "**who** is able to keep that which is committed **unto him.**" (2 Tim., i., 12.)

Such being the gracious gift which it hath pleased God to impart in the Christian dispensation, it would seem mere supererogation **to** handle any of the enigmas of human life as seeking to explain them by human reason. And the admonition of Bacon may here apply : " By aspiring to be like God in power, the angels transgressed and fell—by aspiring to be like God in knowledge, man transgressed and fell ;" and certainly the inquiry into the plan and purpose of Divine Providence, in the affairs of this world, when we know not their relation with the past **or** future, is **full of peril.** But the object I had in view was different. It was rather **to** fix the proper limit of inquiry, by showing **the** perplexity that must still remain after exhausting **all** the arguments of human reason, and having in view chiefly **the** advantage of those who presumptuously set úp *reason* as being adequate to resolve all questions. The self-sufficiency of the worldly mind will slightly heed evangelical truth until its own airy fabrics are **shown to** be entirely without basis.

XXIX

NEUROMATHY.*

THE nervous system has been a late discovery of anatomists. The diseases of **that** system, which are now admitted to be the most formidable that physicians are called to treat, were formerly **sup**posed to be mere old maidish whims, and **were de**signated by the **expressive term of** *Hypo* or *Spleen,* the first of which terms was a **mere** abbreviation of " Hypochondrium," the epigastric region where the spleen is situated ; and the reference of both of the terms being to an organ the office of which is, even to this time, entirely unknown, and therefore supposed to have no function **at** all. There **is some** practical wisdom, therefore, in **the** popular idea that an indolent man is splenetic, **that is to say, that** having no energy of character, he **is** visionary—without purpose—and occupied with thoughts of no actual value.

On the other **hand,** the developments of mental action by preternatural nervous influence, have always been, by a strange inconsistency, **looked**

---

* From νευρον (nerve) and μαθεια (knowledge.) I use the term for convenience in expressing the modern **science of the nervous system.**

upon as traits of genius—or, if a religious element was intermingled, there was a superstitious credence of divine illumination. Persons disordered in mind were supposed, by the Romans, to have the power of presaging future events.*

Witches have invariably had the endowment of a crazed mind. It was necessary, however, to have a fierce insanity—but imbecility would do, if it was of the *maudlin* kind. To be old and hag-like, and rickety of body and mind, was, in the days of witchcraft, sufficient evidence of a communication with spirits, though, of course, it was supposed to be with the evil sort.

I have mentioned that the nervous part of the human organism was, until a recent period, comparatively latent. The muscular bodies of our ancestors, hardened by exercise and exposure, had little of the sensibility which belongs to the refined corporeal structure of the present time. *Now*, the incessant stir of the mind, the habitual excitement in which we live corresponding with the railroad speed of travelling, and the transmission of thought by telegraph, have brought about an *exterior* development of the nerves ; a higher degree of sensibility is diffused over the whole superfices of the

---

* They were called "Ceriti," because *Ceres* was **supposed** to deprive her worshippers of their reason (*Hor. Sat.*, lib, ii. s. 3, 378); sometimes "Larvati" (*Larvarum pieni*—that is, disturbed by communications with ghosts—*larvis et spectris externiti*); sometimes "Lymphatici" (*Ae.* vii., 377 ; *Liv.* vii., 17), from the supposition of their having been the effigies of water nymphs.

body, and a susceptibility, formerly unknown, **now** belongs to the whole interior nervous ramification. It is manifest, therefore, that there is greater facility of receiving external impressions—greater mobility **of feeling** and flexibility **of** purpose—more versatile thought and emotion, but superficial, and lacking consistency and intensity.

It follows that whatever stimulates nervous action acts now *Homœopathically, i. e.,* in infinitesimal proportion.* Alcohol, Opium, and Tobacco, are the principal artificial stimulants of the nerves. *Stramonium, Hyosciamus,* and some other vegetable poisons, have **a** powerful effect, but **they** are used only as medicines.

Indian Hemp is largely used in Asiatic countries and in Egypt. Marvellous effects from its use are reported, but it is likely that the peculiar habits in that part of the world may have much to do therewith. The European constitution does not **appear** to be operated on in the same degree ; **at** any rate, the use of this narcotic has not prevailed much, north of the Mediterranean.

The common hemp of this country has narcotic properties, but the article above mentioned is a wild hemp, which is found largely in India and Egypt, **and** the provinces in the Levant. The scientific

---

* I use the term in a popular sense. Etymologically, Homœopathy has no reference to *quantity* of medicine, but only to the *kind*—the principle being that disease is to be overcome by aggravation, as a crying child is overcome by the louder **vociferation of the nurse.**

name is *Cannabis*. In Asia Minor it is called *Banque*, in Egypt *Assis*, whence is derived the term *Hashish*, by which it has become more generally known. It seems probable that it was known and used in Phrygia by the heathen priests.

All **peculiarities of the** physical organism are hereditary. **The ghastly** aspect of the inebriate **will** not be more surely reproduced in the pallid and woe-stricken faces of his children, than the nervous susceptibility of **the parent** will have **a** congenital development in his offspring. How much of suffering, how much also of the illumination, the lawless imaginations, the spiritual visions, **which** attract so much popular regard, is due to ancestral vice, would **be a** curious subject of inquiry.

I shall **not** attempt any elaborate analysis, but merely refer **to some** striking illustrations of the proposition I have referred **to**.

Von Helmont, who **lived** long **before the supposed** modern device **of** producing mesmeric sleep, was, in fact, the inventor of the whole contrivance —but he is chiefly worthy of recollection from the **fact** that he once fell **into** a spiritual state, and reported that he saw **his own** soul seated in his belly. We **learn, from his own** narrative, that this was the **effect of a large dose of** *Aconite*.

The *Pythonesses* **of** the ancient oracles, as we are now authorized to believe, derived their inspiration from the use of narcotic substances, either eaten in a crude state or inhaled in a gaseous form. Phrygia, which furnished a large proportion of heathen priests and priestesses, was celebrated for its poisons, and for the skill in the use of them which was possessed by those who were devoted to the service of the gods.

That sort of knowledge has, in fact, in every age, belonged to *Sagas, Sorcerers*, and others pretending to inspiration. Even Christianity, at an early period, was somewhat involved in the mysticism connected with this art. The same development that has been referred to as peculiar to the heathen priesthood, was exhibited in some of the early Christians residing in Phrygia. Thus, Montanus set himself up for the Holy Ghost, and his followers, among whom was the celebrated Tertullian, believed that they were themselves divinely illuminated. As the sect was of Phrygian origin, it may be presumed that the illumination was similar to what had existed under the heathen regime.

It is interesting, in this connection, to observe the distinction made by the Greeks between the ψυψη and the πνεμα—the animal soul and spirit— the former of which is designated by the apostle Paul as *earthly*, and by the apostle James as *sensual.**

In speculating upon the influence of one mind

---

over another, it should be noticed that the power
is much greater for a bad purpose than a good one
—there being a natural proclivity of the soul to
evil.

The supposed inspiration which, according to the
theory I have suggested, has its origin in a disor-
ordered nervous system, has, nevertheless, something of grandeur.

Even the heathen Sybil of Cumæ assumed a divine sublimity, and denounced all **profane** famili-
arity with the awful secrets of the spirit world—

" Procul O procul este profani,"

was her language, in the narrative of Virgil.[*]

When Luther was asked how the pretensions of
the **prophets of** Zwickhau to inspiration could be
tested, **his answer** was, " **Ask them** if they have
known **those** heavings **of the soul, those pangs of**
the new creation, those deaths **and hells which be-**
long to regeneration. It behooved our Lord, **through
the** sufferings of death, to enter into glory, **and so
must** every true believer, through the tribulation
of his sins, attain unto rest."

I proceed, however, to the treatment more in detail of the physical process whereby the **nerves** are
acted upon by **external** influences, **the effect of**
which is transmitted to the heart **or** brain, produ-

---

[*] Æ., lib. 6.

cing upon the former an accelerated but irregular *systole* and *diastole*, and upon the latter a *coma*, which may, perhaps, be deemed a congestion rather than what was formerly supposed an illuminated vigil.

I have referred to the power of narcotic stimulants. There is still another element, which is of modern origin, and has a larger operation in this country. It is a common-place remark, that our habit of life has become intensified by the eager pursuit of business. This activity has been set down, by some writers, to the effect of our climate, but it is an imperfect solution ; for, 1st, the aborigines were unemotional, that is, were habitually self-possessed, even stolid, as we would say of men of our own race who are subject to no quick impulses, and are unaffected by social impulse ; 2d, the restlessness displayed by our people has been also seen in later years, to some extent, in the older nations of Europe, particularly those with whom we have much commercial intercourse, viz., England, France, and Germany.

Making, however, due allowance for these considerations, it still may perhaps be true, that our climate *has*, by its sharp alternations, a considerable effect upon the physical system of our people. A more important influence may, however, be referred to the unchecked freedom allowed in this country to every man of pursuing whatever business he chooses,

and the great remuneration which is offered to persevering enterprise. It is possible for a man in very
humble life to attain to the highest public position.
Wealth is the great power ; it is also more generally attainable than any other means of influence,
at least is more open to the general mass of the people. For professional eminence some advantages
of education are required, but in the pursuit of
wealth all that is necessary is industry and practical
sense.

This, then, is a principle deeply implanted in our
entire population ; its motive power was at first
moderate, but it has been constantly accelerated by
the strife of competition, the constantly increasing
momentum of pressure from foreign emigration, and
the rapidly multiplying indigenous population,
and by the encouragement which individual success
in obtaining almost fabulous riches, holds out to the
common mind. To this has been recently added
the vastly increased facilities of trade, telegraphic
communication, railroad transit, &c. Changes in
market now occur with a rapidity that requires incessant watchfulness. The same causes affect all
branches of business ; we are put into a hurry by
seeing others in haste. In addition to this, it is my
opinion that an undue proportion of our population
has been thrown into trade and the professions—
perhaps I should say into the *profession of law.*
The number of ministers and doctors must be limi-

ted by the public demand, but lawyers can make business—and, besides, **they** almost universally go into political affairs ; the pursuit of office is, indeed, peculiar **to the** legal **profession.** Mr. Jefferson, though himself bred **a lawyer,** had a low opinion of the utility of lawyers in public bodies, of which (at least of legislatures) they constitute a large proportion.

I have said, however, enough upon the principle involved. The general habit in this country is an overtasking of body and mind. The **merchant** works from an early to **a** late **hour,** and **takes** his cares and anxieties with **him** when **he** returns home. **If he** travels **he will go** at night, in order **to** save time ; in fact, it is a custom of the greater part **of** our peo-**ple** that travelling by railroad is, for the most part, **done at** night. Even when recreation is the object, **great** distances are travelled without respite ; the night train is often resorted to, sometimes for the sake of economy, but more frequently on account of hurry, **when,** in fact, **there** is nothing requiring dis-patch.

It **is not** my purpose **to** illustrate at any consid-erable length a political **or** social question. I have **to do only** with a physiological fact. As Shake-speare says, in the mock play in Hamlet, in which Pyrrhus is described—

> " Head to foot
> Now is he total gules,"

so **may we** say **of** these **hurried** travellers, that

their nervous system becomes preternaturally developed ; in fact, by constant pressure and over action, the nerves become more exterior—a process like exposing to the open air the delicate and recondite parts of our system. The immediate subject of illustration is, however, the inward effect of the breaking up or impairing the outward sense. It is quite obvious that the surge of human population must often dash upon shallow, or upon rocky shores, and make wreck of many adventurers. My office is only to take notice of this great commotion of the social elements—the rush and refluence of human tides. There is incessant change ; indeed, nothing can be said to be constant. Even the wealth that is accumulated by a long life of successful enterprise is speedily dissipated by the next generation. There seems to be no further development to be looked for in active enterprise—it has already passed beyond the limit within which health and comfort are circumscribed.

According to the witty sketch entitled "The Year Three Thousand,"* muscular development is represented as being effected with reference to a single purpose, being in accordance with the received axiom as to division of labor—thus a blacksmith being gigantic in arms but shrivelled in legs, while dancers have an expansion of legs but diminished chest and arms. So our nervous susceptibility has,

---

* " Harpers' Magazine."

as it were, brought out that part of our interior organism which was **till recently** unknown to physiologists, to such **a** degree that it has become subject to every " skyey influence," and makes up a large proportion of the misery to **which the** civilized part of our race is subject.

XXX

INSPIRATION.

HOW DIVINE KNOWLEDGE IS IMPARTED——WHAT COM-
MUNICATION MAY THERE BE WITH BEINGS OF A
HIGHER ORDER THAN THE HUMAN.

———

RELIGIOUS faith has become, at the present day,
so much more firm and rational than it has been in
former . times, that I have no hesitation in discussing a proposition, which is involved in every Christian creed, with the same candor and fearlessness of
the result that I would have in the investigation of
any psychological theory. It may obviate any
timid apprehension of danger, from this unre-
served mode of dealing with a question heretofore
deemed sacred and mysterious, and far removed
from human inquiry, by stating in the inception
of this brief argument, that I have, to the extent of my ability, sounded all the depths of skep-
ticism, and have adventured upon inquiry into all
that has been suggested by the opponents of our
faith, so far as they have been brought to my knowl-

edge by diligent reading during many years, and
that the result has been an entire confirmation of
my early belief, though the grounds upon which **it**
rests are quite different from those which **once ap-**
peared to me sufficient. **I cannot** but think, therefore, that I shall render no disservice **to** the faith
which I venerate if I state tersely the mode **of** argument by which I have solved, satisfactorily **to** myself, the question involved in the inquiries **I** have
referred to.

Inspiration may **be either direct dictation of**
language, **or it may be the effect of deep devotion
and** intimate communion **with God in** thought. **In**
the latter sense, which I take to be the true hypothesis, we have good authority in respect to all the
sacred writers for considering their doctrines and
precepts as emanating from God himself ; but we
have the right to repose with greater faith upon
some than upon others, according to the measure of
divine influence by which they seem to be actuated,
and this we must judge **of by** evidence appreciable
by human reason.

As to *facts* not essential to the moral truth inculcated, we may suppose them to be mere illustrations
by the inspired writer, derived from his own knowledge, and not more inspired than the *words* that he
uses.

The apostle indeed says, that "all Scripture is
given by inspiration of God" (2 Tim., iii, 16), but

12

this cannot mean that all was *equally* inspired, or that everything contained in them was so.*

Again, it was said, " there is a spirit in *man*, and the Almighty giveth him understanding " (Job, xxxii, 8.) This plainly is an inspiration of which all men in some degree partake. We may conceive the sacred writers to be more largely endowed by the same sort of inspiration but yet subject to human error ; or, in other words, that **they** were more highly gifted with religious perception, but not more than other men with natural or merely human knowledge.

As commonly understood, *inspiration* is an internal communication, not derived through the ordinary modes of acquiring knowledge—that is to say, not from sight or hearing, or other senses, but that it is an impression upon the mind by some process independent of all bodily or sensorial agency—and in this respect, being the same as it would be if the soul were entirely disconnected with the body.

It is true, that what is expressed by the inspired writers is often represented as having been spoken **by** God himself, but I take the true construction to be, when this form is used, that it is merely adopting a mode of speech whereby God is supposed to

---

* I have quoted **from the common** version, but it admits of no doubt that the proper translation is, "all inspired Scripture is profitable," &c. ; referring to the Scriptures of the Old Testament, which were received as inspired. The fair construction of the text referred to is, that the books then received as divine (which are also included in our canon) were inspired.

say directly what has, in fact, been only communicated to the writer. The question still recurs, how did he receive such communication, and it will be difficult to answer it, except upon the assumption above referred to, viz., a direct inspiration, without the intervention of speech.

An objection to this assumption may occur, that there is a want of a reliable test as to the truth of any impression so received. It may be said, that we can communicate with mind only through the medium of language, or expression by the face, or some exterior action. Is it not necessary that God should communicate with us in the same way? He can, indeed, miraculously create faculties that we have not—but that would be to make a new being differing from the human, and the thoughts of such a being would be incomprehensible to others not endowed in like manner, or if they could be comprehended, .what test should we have, such as we apply in judging of all other communicated knowledge?

The evidence in such cases would be, perhaps, no greater than of what occurs in a vision or dream,*

---

* The objection is thus stated by Hobbes: "If a man says that God hath spoken to him supernaturally, I cannot perceive what argument he can produce to make me believe it.   *   *   *   To say that He hath spoken to him in a dream, is no more than to say that he dreamed that God spoke to him, which is not of force to win belief from any man that knows that dreams are, for the most part, natural, and may proceed from former thoughts."—*Leviathan*, chap. 32.

I have no fear of quoting from this redoubted *skeptic* (or, perhaps I should say *heretic*.) It is no more than an objection, fairly stated, which I think I have sufficiently answered.

which, however strongly it may appear to him upon whom the impression is made, as a reality, yet has not the same effect upon one to whom it shall be repeated, because it cannot be sufficiently tested by the ordinary rules of human evidence.

Sharon Turner has made some observations upon this subject that are worthy of consideration: " How can the Omnipotent make himself known to us ? could it be by sight ? clearly not ! He might assume a form, as the Jupiter of antiquity is said to have done, but that would give us no true idea of the Deity. That could only be a temporary assumption of a figure which could have no more to do with his reality than the Egyptian Apis, or Phidian statue, which he was believed formerly to reside in. All visual configuration could but be disguises of himself. ❋ ❋ No visual appearance of the Deity could convey to us his thoughts or will. A voice must express these to us before we could be conscious of them, and this voice must resemble ours, and utter the same vocal sounds which we use to each other, and in the same meanings and phrases—that is, the Deity must, for the time, assume human language, and speak in that style, and in those terms which we are familiar with, and address us like a fellow human being, or he would be unintelligible."❋

This hypothesis is clearly erroneous in one par-

❋ " Turner's Sac. His.," ii, 420.

ticular. Although our ordinary mode of acquiring knowledge is by means of language, or
some physiognomic expression or bodily action,
visible to the eye, yet it is certainly not the only
source. Thought may take a vast range beyond
what has been **thus** communicated. It has been
said that **it** is limited **to** combinations of what
has been seen or heard. It may be **so** in respect to
material objects, but it is clearly **not** so in respect
to abstract truth, or, to speak more precisely, what
is perceived by **the** intellect without the intervention of any **exterior** sense ; and this includes **all**
moral and religious subjects. How many thoughts
come to us without any apparent connection with
the external, or even with what had previously
occupied the mind ; how many suggestions for
good or evil, the origin of which we know not,
and which, therefore, we almost instinctively attribute to some spiritual agency.

What mode of communication there may **be with**
beings not **having a visible presence,** nor using
bodily organs, **we** know not. **That** there must be,
is, however, **certain, as we** ourselves at death must
cease **to** use corporeal functions, at least not the
same that we have been accustomed to here. Again,
an interesting inquiry, which, perhaps, will forever
remain unsolved in this **life, is,** what is **the state**
of the soul in sleep, or apparent unconsciousness,
caused by disease **or bodily injury ? The** rapidity

of thought is such that we may account for dreams by supposing them to have occurred while we are in a half-waking condition, or just as we are coming out of sleep. If this be so, how is the soul occupied in the long intervening sound sleep ? Is there a suspension of its functions, like that of the bodily senses ? This, however, is to state a comparison inaccurately, for there are no bodily powers or sensations but what are derived from, and belong to, the soul, or, at least, cannot exist without its agency. All that we know is, that so far as respects voluntary bodily functions there is a suspension or quiescence ; but what, in the meantime, has become of those powers that are not dependent on corporeal action ?* May it be that in this rest of the earthly frame the intellectual part of our being is left at liberty, **and** that we have thoughts and emotions which are not remembered in our waking hours ? We can hardly suppose an entire absence of the soul from the body, for then life would cease, but may there not be an unknown power, not circumscribed by the limit of corporeal life, whereby the soul in the sleep or inanition of the body may have communion with things never perceived by bodily sense, and **of** which the corporeal spirit (by which I mean **the soul in** its duality of sensuous and intellectual action) is wholly oblivious ? There would be nothing in this more strange than would

---

* I state this in a comparative sense. We are, as yet, imperfectly informed of the exact connection of the mind with the brain and nerves.

be the pre-existence of the soul, and our having no memory of it, and yet there are, at times, some seeming reminiscences, fragmentary and inexplicable, inspiring within us emotions not habitual, as if of association with something that we have not seen or heard in this life, that afford some presumption that the idea of an existence before the present life may turn out to be true.

A common superstition has prevailed in all ages, whereby supernatural beings have been represented as appearing in a human form, yet having only a shadowy outline without substance. Yet have these supposed apparitions been always the occasion of dread. The idea of being brought into intercourse with a spirit inspires a feeling of apprehension totally different from the impression produced by the presence of a man, however great may be his intellectual capacity, or other power, or whatever may be his moral character. Our fellow beings, even the worst of them, have still something in common with us, which preserves a certain degree of likeness in thought and feeling, however much they **may** be perverted ; but when any one has believed in the presence of a being not human, there has been an indescribable awe — a thrill through all the senses.

In the disturbed visions of those persons whose natural reason has been shaken by disorder of the nerves, supernatural appearances **are** easily brought before the mind, but there is trembling even at these fictitious creations of the mind. And the fear consequent upon such imaginations is not wholly divested, if the sufferer should suspect, or even believe, at other times, that they were mere illusions.

With such natural revulsion at the supernatural, it may be difficult to explain how there can be direct intercourse with **God, at least** in the sense commonly understood, without a sense of overwhelming awe. And this, in fact, is expressed by some of the greatest of the prophets.* Yet, however it may **be** explained, it has been, in fact, illustrated in the lives of Christians in former times, and we may form some idea of it from the profound conviction we often see developed in the pious of our own age. This may enable us to realize what was the inspiration imparted to prophets and apostles. No transformation of character by the mere discipline of life, or by culture of the mind, **or by** change of position in the world, can be compared with that which **is** produced in the soul by what is termed in the Scriptures the new birth, or if explained in other **language**, the change of our nature by the mysterious power which works in us when we are brought into communion with God. The life of a man in whom

* Isaiah vi., 5: "Then said I, Woe is me! for I am undone; because I am a man of unclean lips, and I dwell in the midst of a people of unclean lips; for mine eyes have seen the King, the Lord of hosts."

Jeremiah iv., 23–26: "I beheld the earth, and lo, it was without form and void; and the heavens, and they had no light.   *   *   I beheld the mountains, and lo, they trembled, and all the hills moved lightly.   *   *   *   I beheld, **and lo, the** fruitful place was a wilderness, and *all the cities* thereof were broken **down at the** presence **of** the Lord."

this change has been wrought, is not free from the trials to which others are subject, but **by** all such trials, spiritual-mindedness or devotion **to God** becomes more and more controlling.

I have often thought of the self-sacrifice of devout **Christians** as something like the call to the prophetic office. **All ordinary** human motives were in the latter case overthrown by a resistless power that took possession of the soul, and thenceforth controlled its thoughts **and** purposes. Even **the timid and** reluctant became, under **this power,** stern exponents of the word of God ; **bold** in the rebuke of wickedness **in** high places ; yet **withal** retaining their **original** simplicity **of** mind and unpretending manner of life.

To the deprecating Jeremiah, who saw in himself no qualification for this great office, and who, in his pathetic language, said, " Ah, Lord God, behold I cannot speak, for I am a child," the answer **was,** " Say not **I** am a child, for thou shalt **go to all that** I shall send thee, **and** whatsoever **I** shall command thee thou **shalt speak."* Should it** be said that this **was but a communing of the** prophet with his **own** mind, it needs no other reply than that such elevation of soul, and such lofty purposes, so nobly carried out, could **have** been no mere natural **determination.** When, afterwards, by his unwavering declaration of what he deemed himself commissioned of God to say, he had alienated his brethren, and

---

* Jer. i., 6-7.

called down the wrath of the rulers and great men ; when suffering persecution and exposed to peril, he still proceeded on his mission, though sinking under the pressure of the burden imposed upon him.°

So we have the express declaration of St. Paul that he preached not by his own will, but by a power which he could not resist—"Necessity is laid upon me ; yea, woe is me if I preach not the gospel."†

And whoever, in any age, have been called to be instrumental of great good to others, have usually had a preparation that seemed a sacrifice of almost everything that the heart naturally seeks, and the like experience is still exhibited even in our own comparatively placid phase of Christian life. They are made to undergo an ordeal as by fire, whereby all merely human hope and sympathy are blasted ; and at first the soul would appear to be left bare and desolate, in a like condition, though in a worthier sense, with those who were "amerced of heaven," and in poetic imagery compared to the scathed forest oaks or mountain pines—

> "With singed tops their stately growth, though bare,
> Stands on the blasted heath."

But now a new principle of life succeeds to that

---

* I know of nothing more touching in the sacred writings than the prayer, or, more properly, the appeal of the prophet : "O, Lord, thou knowest—remember and visit me.   *   *   Know that for thy sake I have suffered rebuke. Thy words were found, and I did eat them, and thy word was unto me the joy and rejoicing of my heart.   *   *   I sat alone, because of thy hand, for thou hast filled me with indignation."—Jer. xv., 15-16.

† 1 Cor. 9.

which has been extinguished. Self-renunciation is followed by an all-controlling consecration to the service designated by the divine will. From such a calling the natural heart would shrink back with a consciousness of inability—human pride is smitten— the vanity of all merely human purposes and re- sources is demonstrated. But now is realized that *" the way of man is not in himself ; that it is not in man that walketh to direct his steps."* *

In this self-sacrifice and stern devotion to a holy purpose, there is a divine influence imparting tran- quility to the soul—a serene fortitude, and a sense of spiritual things not known in the ordinary expe- rience of life. All the enjoyments of the world, all the objects of worldly ambition, become vapid and unalluring, and in their place is the more satisfy- ing enjoyment of devout meditation—of a clear perception of divine truth, and a consciousness of power to accomplish a noble work, in awakening the consciences of other men, " warning the unruly, comforting the feeble-minded,"† and setting forth, by a life of purity and self-negation, an example that **shall** enlighten those who desire to become " partakers of the divine nature," and rebuke, with silent admonition, the low propensities of the worldly- minded. Such was the apostolic office ; such has been the preparation of many who have been called to testify for the truth before a godless world.

---

* Jer. x., 23.   † Thess. iii., 14.

If it should be asked, **then, what** evidence is there **that** this transformation is by **a divine power,** and how far are we to rely upon what shall be **communicated** by one **who** has thus been isolated, as it were, from his kind ? I answer, that " whatsoever things are pure, whatsoever things are lovely, what**soever things** are of good **report,**[*] commend them**selves to** the consciences of all men as being divine ; **and** wherever these are exhibited, with patient persistency, in the midst of trial, and without any **of** the ordinary incentives by which a Pharasaic pretension to purity of life might **be** prompted, we know that something more than mere human motive **is involved.** When our Saviour proclaimed to the world a law of righteousness, that was to rule the heart as well as the outward actions of men, there was, and ever since has been, a recognition **of His** precepts, as far above all **human speculation. Even** scoffers at the doctrinal religion **of the** Christian **Church** have admitted it. So, whenever the principles **He** taught have been reproduced in the lives of eminently pious men, the world is impressed by the example, and looks upon it with secret respect, though the evil passions of men may be roused to hostility by the rebuke perpetually administered by unselfish purity of life and devotion **to God.**

[*]Phil. iv., 8.

XXXI

NEMESIS; OR, THE RETRIBUTIONS OF LIFE.

ANCIENT SPECULATIONS UPON NATURAL EVIL— LATER HYPOTHESES BASED UPON THE SUPPOSED SUFFICIENCY OF HUMAN REASON, WITHOUT DIVINE REVELATION.

IT is a trite remark, that this life is a probation, by which is meant, that the events of life are designed as a discipline for the restraint of evil propensities, and the nurture of whatever of good there is in our nature.

That there are checks upon the natural predispositions to evil, in every form of society, is, doubtless, true ; but I doubt if all the multiform experience of individual life has, *in itself*, a renovating power. Like the voice of the inward monitor of the soul, it seems rather to be a guard against utter lawlessness, than the vitalizing agency for the progressive advancement of human society. It would, indeed, conflict with all past history, as well as with religious faith, if we were to attribute to the interior working of any merely human agency, the power of elevating and perfecting the character of man.

The civilization of the Greeks and Romans had little effect on the moral tone of character. Vice was not restrained by *virtuous principle*, but only by conflicting passions and interests, which mutually checked each other. Power was the great object of ambition. Qualities essential to the attainment of that object—eloquence, military prowess, political tact, were held in disproportionate esteem. To command armies, or to be at the head of the civil government, was not, however, the only form of power that excited cupidity. Authority over the minds of men was, perhaps, the most attractive object of ambition. To be admired—to be pointed out in the public assembly—to receive public testimonials of honor—may have been more gratifying to human pride than to have control over men by an armed force. But while in the pursuit of power or fame, masculine energies of body and mind were developed, there was no discipline of the moral affections.* Some philosophers did, indeed, commend the *humbler virtues*, but the motive appealed to

---

* **Cicero,** the most humane of ancient moralists, could find no better motive for **the highest virtues,** than **an** intrinsic dignity, which commanded the respect of men.

Thus, to exhibit any virtue which is rare, or to abstain from any vice which it is difficult to resist, was, in his view, the object of special admiration. Pleasures are powerful with most men to turn **them aside from** virtue—he who can despise them exhibits a peculiar splendor. **Life, death,** riches, poverty, vehemently move the minds of men—which things, when a man of lofty mind looks upon unmoved, who does not admire the dignity and beauty of virtue thus exhibited.—*Off., lib.* **2,** § 10.

Such is the reasoning of Cicero. By the same rule of public admiration, he held

was still to gain admiration. It was merely presenting, as a standard, the opinion of the more cultivated class, instead of **that** of the common people.

There certainly was an intellectual perception of what was **right** in itself, as may be seen in the writings of Plato and Cicero — but there was an absence of any sufficient sanction to produce conformity therewith in the life. The applause of mankind, or a particular class of men, was the great inducement held up **to** view by moralists. Virtues relating to private life were little noticed in philosophical speculations, and not at all practised, except so far as the exigencies of society demanded, that is to say, so far as they were enforced, by the laws of the State, for the mutual protection of the citizens. The existence of such laws might, indeed, be deemed an argument in favor of the moral sensibility of the people ; but it will be found that the laws did not extend beyond what was necessary to secure the right of individuals to enjoy those things that they **were impelled to seek** by natural instincts, **so** far **as such** pursuit by each was consistent with that of others.

The laws were imperfect, **even in** this respect— **but** they, **at** all events, did not extend beyond what

---

that is was necessary for *Cato* to kill himself, though it was not for the others who surrendered themselves to Cæsar, for the reason that *he* had acquired renown for indomitable greatness of mind, wherefore it would have impaired the consistency of his life if he had been compelled to yield.—*Off.*, *lib.* 1, § 31.

was essential to the maintenance of society. Perhaps their office may be fairly considered as limited to the imposing of such restraints, as, by common consent, are admitted in lieu of the self-protective right which each individual would have against the aggression of another, and as having no relation to what affected the individual without interfering with the rights of others. The disposition of the heart cannot be reached by laws ; nor, in heathen civilization, were those virtues inculcated at all which spring from social affinity and sympathy. Kindness to neighbors, charity to the poor, a desire for the amelioration of the hardships of life, incident to ignorance and poverty, had no place in Greek or Roman laws, or systems of philosophy.*

---

* The distribution of **bread to the populace** of *Rome*, had its origin in mere political motives. Aspirants **for public favor** competed with each other in procuring largess to the common **people, and** afterwards it became necessary to continue it, in order to prevent insubordination of **an** idle and vicious population.

There was a law of *Thebes* requiring poor parents, who were unable to support their children, to surrender them to the State, which undertook to be at the charge of bringing them up. The object of the law, probably, was to prevent depopulation by infanticide—which, however, it did not accomplish.

The exposure of infants was a general practice in all the Greek cities, and was not repugnant to public sentiment. Even *Plutarch* could excuse it upon moral considerations ("*Plutarch* on the Love of Progeny") ; and the treatment of the **children committed to the State, in the city of Thebes**, proves that a humane interest **in their welfare had but little to do** with the law, as these children could be **sold as slaves for payment of the** charges incurred in supporting them.

At *Sparta*, weak and deformed children were, by law, required to be destroyed, which was done by casting them **into a cavern near Mount** Taygetus. So general was the practice in Greece, that the different **modes** of exposure were expressed by discriminative terms. The place where children were cast was called Ἀποθεσιαι —whence ἀποτιθεναι, to expose with a design to destroy ; ἐκτιθεναι, to expose without a destructive intent ; and both were distinguished from violent taking of life, ἀποκτειναι.

The greatest of the Romans (Julius Cæsar) was the most profligate man of his time. *Cato* the Censor, who was a model of virtue, **according to the** Roman code, **was in the** habit **of getting rid of old** and infirm slaves, **at** any price, **instead of** providing for them himself, and it **was a common** practice, in what was called the *virtuous* **age of the** Roman Republic, to expose slaves, **who were worn out by dis-** ease or the infirmities **of age, on an island in the** Tiber, to perish.*

In this state **of society, what was** really **the effect** of the course **of life upon a man's character ?** Did he become better, in **a moral sense, by all** that experience which **we** term discipline, **or** did he, on the contrary, become harder and more selfish the more **he devoted** himself to the pursuits **of** ambition ?

Another, and a much larger class, to whom **the** path of honor offered little allurement, **either for** want of natural ability, or of other advantages, **we** should presume would be wholly given **up to** sensuality. The intellectual refinement **of the** Athenian might find **gratification in the** drama, in the eloquence **of** popular speakers, and even in the discussions **of** philosophers — the Roman **was** grossly licentious in his pleasures.

Look at another phase of human society, when less of civilization was developed, in **the** Asiatic

---

* Concubinage was entirely unchecked—the purity of domestic life was shamelessly violated by the intercourse of masters with **female slaves.**

nations of a far back period. I refer to that period because, in later times, they have felt the influence, if not of the Christian religion directly, at least of intercourse with Christian nations, which has materially modified their character. It would seem to have been an impossibility, that without religion, and without intellectual cultivation, society could have been held together at all. There must have been some unseen influence, which we do not take into account, whereby mankind, in the lawlessness of their heathen state, were withheld from utter and irretrievable degradation—unfit for the accomplishment of any purpose for which life may be supposed to have been given to man. What was the individual life ? what lesson was there for the advancement of any good principle in the character ? A life spent in vice may bring, at its close, a melancholy retrospect for contemplation—but, if moral sense was not wholly extinct, if conscience had still some power to disturb the soul,* the refuge was in an abject superstition—indeed, at all times, in the midst even of sensual indulgence, there was a fear of some unseen power—a convic-

---

* Conscience had, undoubtedly, some power, even in the most degraded state of human society—to what extent cannot be determined, but its general influence, antagonistic to vice, is certainly to be taken into account.

"Let it be imagined how many men have wished they might be rid of it—let it be imagined with how many men it has interfered to disturb and oppose them • * * Perhaps in no case this could be wholly without effect. The infinite multitude of criminals would have been more criminal but for this."— *Foster's Lectures at Broadmead Chapel.*

tion that there were beings, hostile to men, whom it was necessary to appease. It did not, however, occur to them, that this was to be done by self-restraint and effort to amend their lives. They were merely persuaded, by the pains of body and the hardships and disappointments of which they had experience, that their was some power superior to man, and unfriendly to him. We can hardly suppose a man living always in the indulgence of evil passions, would learn anything good—yet, as before remarked, there is something in the moral agencies by **which** the world is controlled, that interposes checks against vice. Perhaps the domestic relations have most influence. A parent, however sensual he may be himself, would seek to restrain his children, if for no other reason than natural affection—for the evil effect of vice he could not help being convinced of by his own experience.

Plutarch* says, " that the greatest restraint upon wicked men is regard for their posterity—it being a part of the moral government of the world, that a man's descendants are made to feel the effect of his conduct. If his character was bad, his children suffer ignominy, unless by their own excellence they shall prove it to be unjust, and then they will have additional honor for having, against the predisposition to evil, supposed to be inherited, and the disadvantage of public prejudice, achieved a good

* **Discourse** " Concerning those whom God is **slow to** punish."

name." Such are the compensations of life. The wickedness of the parent entails dishonor upon the children—yet a motive is furnished for extraordinary effort by the latter to overcome the bad opinion of the world. And this is an argument of great moment for solving the question, which has perplexed devout men, who have had the light of revelation. The authors of the books of Job and Ecclesiastes seem to have had some idea of this compensation in God's final administration, yet did not fully set forth the effect, on the wicked, of their anxiety for the welfare of their children and friends.

Plutarch, in this respect, had a larger observation, for he insists that retribution does take place by the remorse a man suffers for the evil that he induces upon those who are dependent upon him. Another restraint that he refers to, is the immediate effect of a man's example. A parent is impelled to admonish his children against vice, but he feels, at the same time, that his counsel is of no avail while he is himself indulging in the very thing which he exhorts them to avoid, and hence, is led to check himself in his vicious course, for their sake, which he would not be by any consequences which might visit him alone, and not affect his posterity.

Yet, with all these restraints, private life was licentious among the Asiatics and Greeks, to a degree that seemed to threaten the extinction of society. Upon the whole, would a man, living in that

condition of things, be in **any** sense morally improved by his experience in life ?   **Can** it be shown
for what good end he has lived **at all,** or, **in other**
words, what advantage **it** has been **to** himself or
others that he **has been** sent into the world **?**

Plutarch insists much **upon** the predominance of
evil in human life, **as a** counteraction to **the** fear of
death, and he quotes **the** opinion of **Aristotle,** that
it was best not to be born at all ; **and, next to that,**
it is better **to** die than to live.*

Epicurus **and his** disciples were **led, by the** same
view of the preponderance **of evil over** good, to
make it **an important part of a** man's happiness not
to fear **death, regarding it as** the end of pain, **that**
is to say, **annihilation. The effect** of this was **to**
constitute **it a** part **of their** philosophy to make **the**
most of sensual pleasure, and to overcome pain **by**
energy of mind ; and it ended in the disbelief **of**
Divine Providence altogether.†

Other philosphers held, **it is true, the doctrine of**
the existence of **the soul after this life, but it was a**

---

* The same sentiment is **expressed** in Ecclesiastes : "Wherefore I praised the
dead, which are already dead, more than the living, which are yet alive. Yea,
better is he than they both which hath not been, who hath not seen the evil work
which is done under the sun."—*Eccl.* iv., 2-3.

There is, indeed, a striking similarity in the views of human life, as expressed
in this book, with those entertained by some sects of the Greek philosophers, particularly the Epicurean.

† According to the most learned of the Romans, the elder Pliny, the prevalent
opinion in his time was, that the affairs of **this world were** directed by decrees,

vague conception, and had but little relation to the conduct of a man here ; or, in other words, there was not a distinct idea of penalty in a future state for misconduct in this. The tortures which were represented as inflicted upon certain criminals, were mere poetical fictions. Obscurity and darkness were the worst evils which the philosophers consigned them to.*

Let us now look at the condition of those who live in a community nominally Christian. The larger number now in such a community are unbelievers ; that is to say, though they are acquainted with the doctrines and ethical principles of religion professed by Christians, yet have no experience of a spiritual power working within them, and exercising a control over their natural inclinations. They are therefore in a state differing from that of the heathen only **so far as** modified by the example and influence of professing Christians. And this, **as re-**

---

which, having been primarily established by God, he never afterward interfered with the actual course of things.

Pliny also denied the future existence of the soul. Among other arguments, he insisted that it could not see, or hear, or feel, without the organs which it had been accustomed to use in the present life.

* The theory **of Plutarch is** somewhat vaguely expressed. The present life is simply **a** visible **state—death** is a return to a latent condition. No discrimination would, in this view, exist between the good and the depraved.

"The birth or generation of individuals gives not any being to them which they had not before, but brought the individual into view ; as also the corruption or death of any creature is not the annihilation or reduction into mere nothing, but merely the sending the dissolved being into an invisible state."—*Plutarch, on Living Concealed.*

spects the exterior of life, is considerable.   It gene-
rates, at least, a formality of decorum—an acquies-
cence in certain observances prescribed by the church,
and a decent conformity to the moral tone of society,
kept up by the Christian part of the community.
But if we sift the inner motives of their actions, and
weigh the influences which affect their desires and
secret course of life, it will be found that this external power operates only to check the excess of selfish
passion, or rather to restrain the open display of
unhallowed thoughts and desires, by which they are
in fact secretly actuated, and so far exercising some
force antagonistic to evil propensities, yet still
without inspiring better motives, or awakening a
holier principle of life.   This is a fair view of the
actual condition of such men, and our question is,
how are they operated upon by the various trials of
life, considered as *probative ?*

I remark, in the first place, that one effect, and
immeasurably of greatest consequence is, **that** these
trials are often the means of awakening the conscience, and producing conviction of sin, which is
the first step in the work of regeneration.   The tendency to this result is indeed the distinctive character of the Christian dispensation.

But, second, when such an effect is not produced,
the discipline is substantially the same as was felt
by the heathen mind.   The same arguments that
were canvassed by the old philosophers, rise spon-

taneously to the irreligious **of our own** times. Worldly-minded men, unimpressed **by** evangelical truth, **have** invariably a melancholy view of human life. **It** may seem otherwise, if we judge of them only by what they appear in the full enjoyment of health and capacity for sensual enjoyment. Then they may **be** hilarious, defiant, profane, without **fear** (at least so **far as we can** judge) **of** the future. The **world is,** in their estimation, a good world enough, and abounding **in** pleasures suited **to their** taste. Such, I say, *may* be their case during the brief efflorescence of animal life ; yet brief it neces-**sarily** is, nor is it allotted to all. Pain of body, disappointment **of** hope, some reverse in life, or fail-ure **to attain** what **has** been sought, **soon** come to darken the **mind,** and though a dauntless face may be presented **to the** world, the solitary thoughts of one who **has had such experience, if** disclosed to us, would probably declare **a** spirit resentful of the hard conditions of life. An exaggerated estimate **of** the **evil to** which life is subject, becomes inevitably his practical theory ; in time it will be the burden **of** his conversation, or if he have ambition as an author, **it will** appear in his writings. **A** man of profligate life will **be found to** belong to the same stye with the Epicureans **of old,** discontented, disgusted per-haps, but seeking **to** banish all thought of the future, or of Divine Providence, finding solace chiefly in the belief that human life is left to itself, and at any rate, that when it ends there will be an end of pain. And so, in his attempt to enlighten the world by his writings, he will insist upon human suffering as an overwhelming answer to the Christian doctrine of the goodness of God. Hume's Dialogue on Natural Religion presents substantially the same view of life that was so much insisted on by the Greek philosophy.*

The confession of Chesterfield, one of the most fortunate of men, as the world estimates good fortune, is a memorable testimony to the hollowness of mere human philosophy.† A more pathetic view of human life can scarcely be conceived than is exhib-

---

* " The first entrance into life gives anguish to the new-born infant, and to its wretched parent. Weakness, impotence, distress, attend each stage of that life, and it is at last finished in agony and horror."—*Hume's Dia. on Nat. Religion*, *part* 10.

One rather novel objection to the constitution of animal life is mentioned by Hume, viz.: the meagre provision made beyond what was absolutely necessary for the preservation of the race, there being, according to his hypothesis, no apparent care or concern for the happiness of the creatures composing it, so that they were sufficiently provided for propagation.

† " I have run the silly round of business and of pleasure, and have done with them all. I have enjoyed all the pleasures of the world, and know their futility, and do not regret their loss. I appraise them at their real value, which is in truth very low, whereas those who have not experienced them always overrate them. * * I look back on all that has passed as one of those romantic dreams which opium commonly produces, and I have no wish to repeat the nauseous dose. I have been as wicked and vain as Solomon, but am now at last able to attest the truth of his reflection, that all is vanity and vexation of spirit. Shall I tell you that I bear this situation with resignation ? No, I bear it because I must, whether I will or no. I think of nothing but killing time, now that it has become my enemy. It is my resolution to sleep in the carriage during the rest of life's journey." * * *

So wrote Chesterfield when he had been overtaken by infirmity, and had the near prospect of the approaching end of life.

ited in the history of a man arrived at old age without having found a single object for which life itself was of any worth, and after all **this** experience **of folly,** looking to nothing in the future of **any** better promise.

XXXII

NEMESIS; OR, RETRIBUTIONS OF LIFE.

[CONTINUED.]

SCRIPTURAL VIEW.

LOOKING at human life, as it appears in the light of Christian experience, I remark, in the first place, that a large number even of nominal Christians, perhaps including some who have been brought under the mysterious power of divine grace, yet, when faith is weak, are practically much the same as **the** merely worldly and unbelieving. They are actuated by the same motives in their daily conduct, and, until by some severe dispensation **of** Providence, quickened into a greater sense of **divine** truth, **in** fact live as do men of **the** world, and to them human life must appear just as inexplicable as it does to the heathen mind.

If we proceed still further, and assume all the knowledge which is derived from the Scriptures and genuine change of heart, let us consider whether, with the aid thus derived, all the difficulty is overcome. The scriptural view appears to be that suffering is a chastisement for **sins, not vindictive, but** for our correction, in which view **human life is** probative, or disciplinary. A question still remains, whether, **with our** imperfect knowledge, we can discern a correspondence in the good accomplished with **the** suffering **by** which it is wrought, or in other **words,** in exact proportion between the **trial** and the benefit.

The following observations naturally **occur in at-** tempting **a** solution of the question : **1. The evils of human** life exist in greatest degree among **those** who are least competent to feel any moral influence therefrom. This, at least, so appears with respect to physical sufferings, and the want of the ordinary comforts of life, whether these relate to the body or the mind. The slave is too much degraded to feel any moral effect from chastisement. The downtrodden people, in countries where there is **a** despotic government, are reduced to a mere animal state of existence. Oppression extinguishes the spiritual nature. A man who has nothing that **he** can call his own, and who can look forward to no improvement of his state, upon whom no genial in- **fluences are shed by the** kindness of his superiors in rank, but only a blight and a curse from the self- ish passions of men to whom **he** is subject, will sink into a brutish sensuality, or cherish in his heart a feeling that he has been unjustly dealt with, and a secret longing for revenge.

The voice of Christian piety may, indeed, reach such a heart. It was among the degraded and outcast that our Saviour was most listened to. "Verily I say unto you that the publicans and harlots go into the kingdom before you," was his stern rebuke to the priests and elders. But how large a part of the human race are out of the reach of the consolations administered by the preaching of the Gospel. Even in those more favored countries where the religion of Christ is professed, and where the government is comparatively mild, what multitudes are living in a benighted state, unvisited by the messages of truth, without Christian association or public worship, or knowledge of the Scriptures. To all who are thus involved in heathenish darkness or in worse than heathenish neglect, in Christian lands, what is there in suffering, unministered to and unalleviated by human charity or divine solace, that is calculated to elevate the thought of the oppressed spirit to God as a merciful being? What self-amelioration is there in their condition? what tendency to purge the soul of brutish passion, and in its stead to foster submission to a divine will? what sense of the goodness of God may we suppose to be generated in the midst of the evils by which their unhappy lot in life is surrounded? These are questions which press painfully upon the thoughtful mind, a satisfactory answer to which is not easily found.

2. Extending our inquiry to the condition of those who are more happily placed in life—who have religious instruction and Christian association—we can see the development of a plan which, under certain circumstances, is potent in its influence upon human character. In the life of a Christian who has lived to full age, all the events of its entire course seem to be so ordered as to bring into action holy principles which are latent in the natural heart, but by divine grace are nurtured into vitality. To such a man the whole of life is progressive ; all his experience has wrought addition to his faith and purity of character. But when we reflect how few are permitted to enjoy this complete accomplishment of a consecutive discipline, the large proportion of those who are but just entering life when they are called from it, a distressing doubt arises, as to the far greater proportion of a Christian community, what to them is the purpose of life ? If for probation, *that*, so far as we can see, is incomplete—or may it be supposed that in another state of existence probation may be continued to such as have been curtailed of it here ? This seems not to be the scriptural doctrine ; yet we are prone to seek some general law to which all men are subject. Perhaps we should look on this as one of those speculative questions which we cannot comprehend without a knowledge of other things which have not been revealed, and, therefore, should not allow it

to disturb our faith or impair our confidence in the goodness or wisdom of God in his arrangement of human affairs.

If mere hypothesis might be allowed, we might almost yield to the suggestion that there is some relation between this earth and other worlds which is now involved in mystery, and that among other things it may be supposed that this is the nursery of the spiritual principle, which is to be more fully developed elsewhere.* This hypothesis would apply to the brute creation also, but that would perhaps form no objection, for I know no positive proof that the animal soul will not survive this life.

3. Looking at the Christian life, and considering the adaptations of all the arrangements of the course of events in the world, with reference to their effect on such a life, and leaving out of view all speculative questions such as those before suggested, there are two views which differ somewhat in their aspect. One is that the life of a Christian is subject to just so much of trial as is necessary to maintain a tender conscience, and a quickened sense of the need of divine grace to resist temptations, and thus to make his life progressive. In this view the ills that we are subject to are like the restraints or the penalties imposed upon chil-

---

* "It is here brought into its first state of being, in animal forms, with a profusion that seems to us unexplainably lavish, unless in order to be used elsewhere in some advanced or ulterior condition, and in other modes of material existence."—*Sharon Turner, Sac. His.*, v. i, 389.

dren by parents and instructors—and human ills might be considered as a process of education. This hypothesis assumes that the tendencies of the soul are not so strongly evil but that they may be kept from perversion by educational discipline. The other view is, that the inclinations of the heart are *wholly* evil and opposed to the law of God, and that they are brought into conformity with that law only by an entire renewal of our nature, which our Saviour compared to a new birth. The latter is undoubtedly the more scriptural view.

Of the origin of this state we know nothing. It may be, as Bishop Butler suggested, connected with something in the past of which we have not the history. The bolder hypothesis of others is not unnatural, though unsustained by proof, that when we enter the spiritual world it may be found that we have a past history preceding the present life.*

---

* Pre-existence was a favorite theory with some of the Greek philosophers. Plutarch refers to it, not as a mere hypothesis, but affirms it as if it were positively proven. Life, in his view, was merely becoming visible, and death was returning to a latent state.

So he says "Apollo (the sun) is called *Delius* and *Pythius*, that is *conspicuous* and *known;* but the ruler of the infernal regions is called *Hades*, that is invisible, and man himself was at first called *Phos* (light.) There being an affinity or perpetual desire in mankind of seeing and being seen by each other."—*Plutarch, on Living Concealed.*

The same idea seems to take hold of the irreligious speculative mind, as expressed by a poet of our own times:

> " I look upon the peopled desert past
> As on a place of agony and strife ;
> Where, for some sin to sorrow I was cast,
> To act and suffer—but remount at last."

Pre-existence cannot be proved, neither can it be disproved, except by the ab-

There is a congruity in human experience with a great **and** beneficent design, if we assume the fallen nature of man, whatever may have been the cause. We see, **then, the** necessity of the many and diverse forms of suffering ; pain **of body,** distress of **mind,** disappointment **of** hope, and all the **sorrows of** which every human being has his share, though in no fixed proportion, so far as **we** are able **to deter-** mine. For this diversity we **may** suppose **a proba-** ble cause, that, although all have sinned, yet **is there** still an inconceivable variety **of** character, **and a** corresponding susceptibility **to** the influence **of the** circumstances affecting human life. This, indeed, **is the** necessary result of the free action of man's will—the various degrees of capacity allotted to the myriad beings of the human **race,** and the infinite **variety of conditions in** which they are placed.

---

sence of proof in its favor. Consciousness or recollection is the ground of belief as **to** the present life, yet the loss of it, as to any **former period** of this life, does not prove that we did not then live. **There** is other evidence—the recollection by others—which we **cannot** have **as to a previous existence.** The most renowned of the Christian Fathers **(Origen) believed that human** souls have existed in an ethe- rial state, with **the power of choosing good** or evil ; that for the voluntary choice of **evil** they have been inclosed **in material bodies.** Some inhabit the stars ; others, whose sin had been more heinous, have become the successive generations of man- **kind** ; and our health and richness, beauty and deformity, prosperity and adversity, are in proportion to our deserts in a former state. All suffering he considered **to be** designed for our renovation, and when we shall have been purged of our sins **we shall** be restored to our first condition. Origen probably derived this **opinion** from oriental philosophy, but Christian sympathy was largely intermingled ; he believed in **the** final restoration of all sinful beings, even Satan himself.—*See article on Origen, Brit. Quar. Rev. (Eclectic, Jan.,* 1846.)

The same stupendous wisdom which is exhibited in the very constitution of the race, by which, in the countless multitudes that have lived, individuality has been maintained, while there has been a nature common to all, we may take for granted is also competent to administer with discriminating precision the affairs of human life, so as to apportion to each individual the peculiar experience which is best adapted to him.

This does not necessarily induce the presumption that those whose trials are apparently greatest are therefore more reprobate in mind than those who seem to have a milder dispensation. Overt suffering is no test what may actually be the trial of the soul. "The heart knoweth its own bitterness" (Prov. xiv., 10), and there are vast numbers of men who are seemingly exempt from the grosser and more obvious pains of life, who are in the enjoyment of what the heart naturally craves as the best condition of life—affluence and high position—who yet would gladly exchange all their fancied happiness for the "weariest and most loathed worldly life" ever imposed on man by penury and bodily pain.

It is enough, perhaps, for us to know that trial of some sort is the lot of every human being, and that it is the only method which, in the moral administration of the world, has been found effectual to restrain the proclivity of man to lawless wickedness, and to reclaim him from the slavery of sin. It will, of course, not be supposed that I am presenting a view of human life as if it consisted merely of penalties, which being suffered, the soul is thereby **restored** from the consequences of transgression. **On** the contrary, I assume that the human soul, in its natural state, before the work of divine **grace** has been wrought therein, is without hope, and that, without a revelation from God himself, no voice of comfort could ever have reached our ears, nor peace have revisited the sin-stricken conscience ; but even with all the divine aid accorded to frail humanity, **it** seems to have been found necessary **for the reno**vation of the soul, that there should be personal experience of some of the results of, or penalties due to, an unhallowed life. We are all, while unrenovated, in the condition of the prodigal son, and not all the yearnings of a father's heart towards **us,** nor all the assurance that he is ready to forgive **our** past transgressions and receive us again to the paternal home, whenever we are willing to renounce **our** life of sin and return to filial obedience, would ever induce us to take that step, if we were not first re-**duced to** feed **on** husks allotted to swine, and feel the bitterness of degradation and despair.

Again would I guard against any possible misconstruction. The ills of life, as the worldly mind estimates them, constitute but a part of the trials which make up the discipline of the soul. Pain of body, the loss of property or friends, and other misfortunes, are calculated to awaken a sense of our condition of dependence upon God, and utter penury of all that the immortal soul should seek. But other causes may arouse the same reflections. Whatever providential dealings may bring about the consciousness of sin, and a feeling of penitence under that conviction, may be deemed a part of the discipline by which, with no more than necessary severity, we are chastened for our good.

But there may be secret wounds of conscience where there is no visible dealing of Providence with us. "As Thy favors have increased upon me, so have Thy corrections. * * Ever as my worldly blessings have been exalted, so secret darts from Thee have pierced me," was the devout confession of Bacon. The outward and physical calamities of life are neither discriminative judgments for sin, nor do they constitute the only means of awakening the conscience. "Those eighteen upon whom the tower of Siloam fell and slew them, think ye that they were sinners above all that dwelt in Jerusalem? I tell you nay, but except ye repent ye shall all likewise perish." These were the words of our Saviour, and we are taught thereby that we all have need of the same salvation, and that none are punished vindictively in this life.

Another hypothesis has been insisted upon, which has probably grown out of the observation of the eminently useful qualities of many men who have had severe trials, viz., that all trials are designed as a preparation for the accomplishment of good to those who suffer, and through their instrumentality to others. According to this theory, great afflictions should be deemed evidence of a special designation of the person so tried for a ministry of more than ordinary import. As it was declared in a vision to Ananias that Paul was "a chosen vessel," and it was added, "I will show him how great things he must suffer for my name's sake," so every one chosen for a work of difficulty may be supposed to have a previous discipline by actual endurance, and a revelation of greater trials to be undergone. There is undoubtedly good reason for believing that this is often the mode in which the most eminent Christian efficiency is wrought out. Self-sacrifice precedes those grand developments of power by which the human mind is moved in great emergencies. This power comes from the renewal of the interior life, the renunciation of what the unregenerate heart desires, and the awakening of noble purposes, which may be thought of with satisfaction in solitary meditation—which can be prayed for, and the very contemplation of, and prayer for which, elevate the soul to a clearer view of divine truth, and greater conformity to that truth in thought and action. This constitutes fellowship with Christ, so much insisted upon by the Apostles: "We glory in you (said Paul to the Thessalonians), for your patience and faith in **all** your persecutions and tribulations which ye endure ; **which is** a manifest token of the righteous judgment of God that **ye** should be counted worthy of the kingdom of God, for which **also** ye suffer."

So to the Philippians, " that I **may know** the fellowship of his sufferings, and **be made** conformable **unto** his death ;" and **to** Timothy, " **if we suffer** we shall **also reign** with **him.**"   So also St. Peter, " that **the trial of** your faith   *   *   *   might be found unto **praise,** and honor, and glory, **at the appearing** of **Jesus** Christ."

It is, however, carrying this doctrine **to** an extreme degree, **to** suppose that the ordinary proceeding **of** the Holy Spirit, in renewing the soul, is by **the disappointment** of all the natural hopes and desires **of the human heart—though** it may be true in **respect to individuals who are** called to some great undertaking, **yet not of all** Christians, **nor** even the larger number.*

In the review of this subject, **there** may, perhaps, **be** left **an** impression that the result arrived at **is** indefinite—something akin to the doubtful deduc-

---

* The following extract expresses the ultra views to which I have referred:

" The word of Providence **and the** Spirit is applied successively to every tie that binds them to the world. **Their** property, their health, their friends, fall before it. The inward fabric of hopes and joys, where self-love was nourished, and pride had its nest, is leveled to the dust. They are smitten within and without—burned with fire—overwhelmed with the waters—peeled, scathed, and blasted to the very extremity of endurance—till they learn, in this dreadful baptism, the inconsistency of the attempted worship of God and mammon at the same time."—*Upham's Interior Life.*

tions of heathen philosophy—there may be, even in
the minds of some who are unaccustomed to pursue
fearlessly the investigation of truth, a vague apprehension of departure from scriptural doctrine. To
such persons the following considerations may be
properly addressed :

I. It is apparent that the entire scheme of human
life, and all its relations to the past, the future, and
the spiritual, have not been fully revealed in the
word of God, perhaps could not have been comprehended by our imperfect faculties if the revelation
had been made.

II. It seems not to have been intended that all
speculative questioning should have a solution,
nor is it known to us why the measure of our knowledge is thus limited. It is, as before suggested, a
probable hypothesis, yet still only conjectural, that
the very difficulty of arriving at absolute certainty
as to many things which we deem essential to our
future welfare, may be for the exercise of our faith.
The importance, as it may appear to us, of having
clear, reliable knowledge upon certain subjects, is,
by no means, a conclusive argument that it is so regarded by divine wisdom ; nor, indeed, is it, according to the analogy of God's dealing with us, in
many things relating to this world, in which we suppose our happiness to be deeply involved.*

* " Would it not have been thought highly improbable that man should have
been so much more capable of discovering, even to a certainty, the general laws of

III. We have, in the experience of life, as well as
in scriptural precept, a demonstration that we have
to do with *practical duties,* and that the culture of
true Christian virtues depends, not upon profound
knowledge, but upon conformity to the example of
our Lord—upon a ready obedience to all that is
enjoined by divine truth, and the inward promptings
of our own consciences, enlightened by the Holy
Spirit, who, we cannot doubt, does abide with every
penitent and believing soul. The lost angels are
well represented as discussing abstract doctrines and
still seeking to explore into their intricate mazes,
though in their efforts they are more and more in-
extricably involved in doubts. So, irreligious men
have ever been the most prone to occupy themselves
with queries having no immediate bearing upon the
conduct of life. In fact, this very habit of mind
grows out of the natural aversion of the human
heart to conform with the plain, practical precepts
for the regulation of life, which are clearly pre-

---

matter, and the magnitudes, paths, and revolutions of the heavenly bodies, than
the occasions and causes of distempers, and many other things in which human
life seems so much more nearly concerned than in astronomy."—*Butler's Analogy,*
p. 2, c. 3.

"We know, indeed, several of the general laws of matter; and a great part of
the natural behaviour of living agents is reducible to general laws. But we know,
in a manner, nothing by what laws storms and tempests, earthquakes, famine,
pestilence, become the instruments of destruction to mankind, and the laws by
which persons born into the world at such a time and place are of such capacities, ge-
niuses, tempers—the laws by which thoughts come into our mind in a multitude of
cases, and by which innumerable things happen of the greatest influence upon the
affairs and state of the world."—*Ib.,* c. 4.

scribed in the Scriptures. It is the part of wisdom to learn what is the true limit of human knowledge, and not obstinately to persist in speculations which can minister only to *unbelief*—in beating the air to no purpose, while there are so many and pressing demands for all our energies in active service.

Lastly. We know with certainty that we are in a state of trial in this life—and though we are not informed of the whole scope of the discipline to which we are subject, nor all the purposes for which it is needful, nor why we see so little uniformity in its administration, still we are sufficiently informed of the *fact*, that in the system of the moral government of this world, we are brought under a discipline. A genuine faith in the great truths of revealed religion should induce us to heed with filial docility whatsoever admonition comes to us in the providence of God, and diligently to seek for that entire conformity of the disposition of the heart to the sublime purity which is brought before us in the life and teaching of Jesus Christ. If we ought to pray that God would sanctify us wholly in spirit, soul, and body, then may we also hope that we shall have a gracious answer to that prayer, and that we shall find rest in God. Yet not the rest of inaction—not the passive rest of cessation from labor and self-watchfulness—not relief from pain of body and other sufferings incident to human life, but the rest of peace—when **we shall** no longer be agitated by fear of what may befall us, but **be** able **to commit our** way unto the Lord and trust in Him. Even our Saviour, it is said, " though he were **a** Son, yet learned he obedience **by the** things which **he** suffered."

www.ingramcontent.com/pod-product-compliance
Lightning Source LLC
Chambersburg PA
CBHW021506210326
41599CB00012B/1149